Science Denial

Science Denial

Why It Happens and What to Do About It

GALE M. SINATRA AND BARBARA K. HOFER

OXFORD
UNIVERSITY PRESS

Oxford University Press is a department of the University of Oxford. It furthers
the University's objective of excellence in research, scholarship, and education
by publishing worldwide. Oxford is a registered trade mark of Oxford University
Press in the UK and certain other countries.

Published in the United States of America by Oxford University Press
198 Madison Avenue, New York, NY 10016, United States of America.

Library of Congress Cataloging-in-Publication Data
Names: Sinatra, Gale M., author. | Hofer, Barbara K., author.
Title: Science denial : why it happens and what to do about it / Gale M. Sinatra,
Barbara K. Hofer.
Description: New York : Oxford University Press, 2021. | Includes
bibliographical references and index.
Identifiers: LCCN 2021004425 (print) | LCCN 2021004426 (ebook) |
ISBN 9780190944681 (hardback) | ISBN 9780190944704 (epub) |
ISBN 9780190944711
Subjects: LCSH: Science—Social aspects—United States. |
Science—Political aspects—United States. | Pseudoscience—United
States. | Disinformation—United States.
Classification: LCC Q175.52.U6 S56 2021 (print) | LCC Q175.52.U6 (ebook)
| DDC 500—dc23
LC record available at https://lccn.loc.gov/2021004425
LC ebook record available at https://lccn.loc.gov/2021004426

DOI: 10.1093/oso/9780190944681.001.0001

5 7 9 8 6

Printed by Sheridan Books, Inc., United States of America

To my nephew Nick, the scientist
GMS

To my grandsons Fox and Leo, budding scientific thinkers
BKH

Contents

Preface   ix

SECTION I: SCIENCE DENIAL, DOUBT, AND RESISTANCE

1. What Is the Problem and Why Does It Matter?   3

2. How Do We Make Sense of Science Claims Online?   23

3. What Role Can Science Education Play?   50

SECTION II: FIVE EXPLANATIONS FOR SCIENCE DENIAL, DOUBT, AND RESISTANCE

4. How Do Cognitive Biases Influence Reasoning?   77

5. How Do Individuals Think About Knowledge and Knowing?   97

6. What Motivates People to Question Science?   122

7. How Do Emotions and Attitudes Influence Science
   Understanding?   142

8. What Can We Do About Science Denial, Doubt,
   and Resistance?   161

Index   185

Preface

Science denial, doubt, and resistance have been persistent and growing problems in the United States, as well as in other countries. At no time, however, have the consequences been as deadly as in 2020, with a global pandemic ravaging the world's population, spreading rapidly when unchecked, fueled by denial of its lethality and the steps needed to contain it. At the same time, the effects of climate change continue unabated, threatening life on this planet. We have never felt more passionate about the need for this book and our desire to help others make sense of the psychological explanations for science denial, as well as learning how to address it.

The day we submitted a draft of this book, February 25, 2020, the US Centers for Disease Control and Prevention (CDC) alerted the nation to the potential community spread of the novel coronavirus. At that time only 14 cases had been documented in the United States, with the first US case confirmed more than a month earlier on January 20.[1] When we drafted this preface in mid-June 2020, the world had exceeded 8 million cases, with a full quarter of the cases—and deaths—in the United States, which has only 4% of the world's population. A reported 22 states were seeing their highest number of positive tests of COVID-19, most likely as a result of reopening restaurants, gyms, and other businesses, in spite of what scientific interpretation of the data suggested. Now, just 9 months later, in March 2021, the world has had 120 million cases, and the United States continues to lead the world in the spread of the virus, even as increasing numbers of people are vaccinated. More than 500,000 in the United States have died. Yet some elected officials have failed to model the most basic protective behaviors of social distancing and mask wearing recommended by the CDC. Throughout these months of the pandemic, science denial has indeed become deadly.

The scientific and medical community warned that the virus was dangerous and that unprecedent steps were immediately necessary to limit the potential catastrophic impact. The United States was slow to act, however, in terms of testing, contact tracing, travel limitations, and orders to stay at home. Scientists quickly worked to make sense of the novel coronavirus and the disease it causes, COVID-19, with an understandable degree of uncertainty in the epidemiological modeling and in the early advice about prevention, with no clear-cut predictions about when vaccines and treatments would exist. Yet their growing knowledge did not translate to policy at the national level. Within weeks of the onset of the virus, a barrage of confusing, misleading, and inaccurate information spread faster than the virus itself. Misinformation was coming from all corners: politicians, newscasts, social media, dark web conspiracy theorists, and even the presidential briefing room. Never has it been so imperative that individuals have the skills to understand scientific inquiry and know how to evaluate what they read and hear. Nor has it ever been more critical that policy makers listen to scientists and make use of scientific data to guide decisions to protect their communities, states, and countries. The first vaccine received emergency approval in mid-December 2020, others have followed, and the rate of vaccinations grows. Yet worries persist about whether enough citizens will choose to be vaccinated, a deeply concerning problem when such decisions affect others' lives, not just one's own.

As the virus has raged, the differential and more damaging effects on those who are Black, Hispanic/Latino and Indigenous have become increasingly evident. Disparities in access to quality healthcare and overrepresentation in jobs deemed "essential" exposed the impact of systemic racism yet again. Racial gaps were evident in all age groups and most pronounced in middle age, with death rates at least 6 times higher for Black and Hispanic/Latino individuals than for Whites.[2] (Such differences also exist for Alaska Native and American Indian groups compared to Whites, but incomplete data prevented analyses at the same level.) For too long, science has ignored issues of structural racism within the discipline and the practice of science. The need for a more socially just science that would examine such inequalities is paramount.

As the virus wreaks havoc on the world and individuals both deny its existence and avoid preventative measures, we are increasingly aware of the need to explore and interpret the conditions for science denial, doubt, and resistance. In this book we shed light on key psychological reasons that have been the subject of our research independently and collaboratively over the years: cognitive biases, evaluating science claims, science knowledge, motivated reasoning, social identity, beliefs about knowledge and science, and the effect of attitudes and emotions. We also offer action steps for individuals, educators, science communicators, and policy makers to support public understanding of science.

The idea for this book came from our individual and collaborative work over two decades. Many long conversations at conferences and meetings about public understanding and misunderstanding of science, coupled with our complementary research agendas in the psychology of thinking and learning about science, led us to collaborate on an article for *Policy Insights from the Behavioral and Brain Sciences*, entitled "Public Understanding of Science: Policy and Educational Implications."[3] That article, others, and many conversations since led us to the desire to speak to a broader audience than those who read psychology research journals.

We extend our gratitude to those who have supported us in writing this book, collaborated with us on related research, and assisted in many other ways. Thank you to our colleagues and collaborators whose work directly informed our thinking in these chapters, especially several of Gale's former graduate students, Jackie Cordova, Robert Danielson, Ben Heddy, Marcus Johnson, Suzanne Broughton Jones, Doug Lombardi, and Louis Nadelson, and at Middlebury, Barbara's undergraduate research assistants, especially Alex DeLisi, Lauren Goldstein, Katie Greis, Amber Harris, Chelsea Jerome, Chak Fu Lam, Jonas Schoenfeld, and Haley Tretault. We appreciate the extensive conversations with colleagues whose work also addresses the public understanding of science: Ivar Bråten, Rainer Bromme, Clark Chinn, Susan Fiske, Heidi Grasswick, Jeff Greene, Susan Goldman, Michelle McCauley, Krista Muis, Michael Ranney, Viviane Seyranian, and Andrew Shtulman. We thank those who read and commented on earlier chapters: Tim Case, Donna Decker, Mike Gorrell, Jennifer Gribben, Joan Sinatra Hathaway, Imogen Herrick, Neil Jacobson, Alana Kennedy, Ann Kim, Ananya Matewos, Beverly McCay, Catherine Nicastro, and Ian Thacker. Our special thanks go to Doug Lombardi, who provided feedback on the entire book and gave it a trial run in his course at the University of Maryland. We are especially appreciative of reviews we received both at the proposal stage and with the final manuscript and are grateful for the time, attention, and critical reading provided by these anonymous reviewers. Family and friends have been gracious in their support and discussions, including Zach, Erin, and Selene Hofer-Shall; and Helen Young, Kirsten Hoving, and Carole Cavanaugh.

We have been inspired in our own research and theory-building on the importance of scientific understanding by the writing and research of others, including Eric Conway, Sara Gorman, Jack Gorman, Naomi Klein, Stephan Lewandowsky, Michael Mann, Lee McIntyre, Bill McKibben, Chris Mooney, Naomi Oreskes, Shawn Otto, Priti Shah, Per Espen Stokenes, and Neil deGrasse Tyson. Other writers who have influenced our thinking about the broader range of topics addressed in this book include Dan Ariely, Philip Fernbach, Michael Patrick Lynch, Tom Nichols, Michael Nussbaum, Eli Pariser, Jennifer Reich, Steven Sloman, Keith Stanovich, and Sam Wineburg. Mistakes and misrepresentations are indeed our own, and we welcome corrections and elaborations.

We appreciate funding received from the Rossier School of Education, University of Southern California (Gale) and Middlebury College, the National Science Foundation, and Vermont EPSCoR (Barbara). We are grateful to our team at Oxford University Press, particularly our editors Joan Bossert and Abby Gross and assistant editors Phil Velinov and Katie Pratt. Finally, most importantly, gratitude to Frank from Gale: thank you for all you do and for your loving support. To Tim, from Barbara: thank you for your own commitments and efforts to address climate change and science denial, for endless conversations on these topics, and for your love and good humor always.

Notes

1. Michelle L. Holshue et al., "First Case of 2019 Novel Coronavirus in the United States," *New England Journal of Medicine* (January 31, 2020), https://www.nejm.org/doi/10.1056/NEJMoa2001191.
2. Tiffany Ford, Sarah Reber, and Richard V. Reeves, "Race Gaps in COVID-19 Deaths Are Even Bigger Than They Appear" (Brookings Institute, June 16, 2020), https://www.brookings.edu/blog/up-front/2020/06/16/race-gaps-in-covid-19-deaths-are-even-bigger-than-they-appear/.
3. Gale M. Sinatra and Barbara K. Hofer, "Public Understanding of Science: Policy and Educational Implications," *Policy Insights from the Behavioral and Brain Sciences* 3, no. 2 (2016).

References

Ford, Tiffany, Sarah Reber, and Richard V. Reeves. "Race Gaps in COVID-19 Deaths Are Even Bigger Than They Appear." *Brookings Institute*, June 16, 2020. https://www.brookings.edu/blog/up-front/2020/06/16/race-gaps-in-covid-19-deaths-are-even-bigger-than-they-appear/.

Holshue, Michelle L., Chas DeBolt, Scott Lindquist, Kathy H. Lofy, John Wiesman, Hollianne Bruce, Christopher Spitters et al. "First Case of 2019 Novel Coronavirus in the United States." *New England Journal of Medicine*, January 31, 2020. https://www. nejm.org/doi/10.1056/NEJMoa2001191.

Sinatra, Gale M., and Barbara K. Hofer. "Public Understanding of Science: Policy and Educational Implications." *Policy Insights from the Behavioral and Brain Sciences* 3, no. 2 (2016): 245–53.

SECTION I

SCIENCE DENIAL, DOUBT, AND RESISTANCE

1

What Is the Problem and Why Does It Matter?

The idea that science should be our dominant source of authority about empirical matters—about matters of fact—is one that has prevailed in Western countries since the Enlightenment, but it can no longer be sustained without an argument.

Naomi Oreskes, *Why Trust Science*[1]

How do individuals decide whether to vaccinate their children against childhood diseases, wear a mask during a pandemic, or eat foods that have been genetically modified? How can they fairly evaluate the environmental and public health risks of fracking or climate change? In a democracy, educated citizens must make informed decisions about scientific issues. However, as people read online news and information or scan social media accounts to connect with friends and family, they are confronted with complex and often conflicting information about science. Evaluating this information is necessary to make consequential decisions that impact one's health and well-being, as well as communities, nations, and the planet. Yet many individuals question or deny the scientific consensus on critical issues or lack the skills to assess media reports of scientific findings. Many do not know or misunderstand the scientific process that produces these findings, challenging their ability to understand and evaluate research results presented in the media. They may not understand or appreciate the role of scientists in contributing to theory, creating advancements in science that are elegant and explanatory.

A gap exists between scientific knowledge and the public understanding and acceptance of science. Over 98% of climate scientists concur that humans are causing climate change, but only 57% of the US public think climate change is mostly caused by human activities.[2] Most parents vaccinate their children; but many do not, or they delay the process or make selective choices,

Science Denial. Gale M. Sinatra and Barbara K. Hofer, Oxford University Press. © Oxford University Press 2021.
DOI: 10.1093/oso/9780190944681.003.0001

putting their children and others at risk. Some question the value of getting vaccinations themselves, for flu or COVID-19. Communities vote to ban public water fluoridation without sound scientific reasons. Consumers buy organic, gluten-free, or non-GMO (genetically modified organism) foods, often without a clear understanding of what these labels mean. No one demographic or political group has a corner on science doubt or denial. Science misunderstanding is evident across racial, gender, age, and political lines.

Perhaps better science education can help address some of the issues of science misunderstanding, doubt, and denial. This does not mean, however, that we simply need to address a knowledge deficit among the public to fix the problem or that such a deficit is the root of the problem.[3,4] Through our own research and that of many others in psychology and education, we have come to understand how making decisions about complex scientific topics requires more than just better knowledge of the facts. It takes the ability to critically evaluate evidence and explanations, take into account the source of that information, and appreciate how the methods of science lead to specific conclusions. People need to know where to turn for reliable information, whom to trust on issues of science, why to value science, and how to resolve conflicting points of view.

Individual actions alone will not address the pressing issues of our era, but understanding and addressing individual resistance and misunderstanding of science can further the potential for the collective action that is needed. Citizens who acknowledge the human causes of climate change, listen to the advice of medical experts during a pandemic, or are able to interpret data regarding gun violence are likely to be far more prepared to support initiatives that will improve health and well-being of the planet and its communities and inhabitants.

Why Value Science?

Since humans began to ponder their own existence, they have wondered about the natural world and their place in it. Why does the sun appear to move across the sky? Why do the seasons change? What causes illness? Prior to the scientific revolution, humans were at the mercy of the elements of nature. Maladies of all types were a mystery with no means of redress other than hope and superstition.

Science has led to remarkable discoveries, such as the eradication of diseases that once claimed lives by the millions, the sequencing of the human genome, and effective treatments for many forms of cancer and HIV/AIDS.

Science provides insights into the origins of our universe, the chemistry of the brain and many mental health disorders, how an asteroid striking the Earth contributed to the extinction of dinosaurs, and how the dramatic rise in CO_2 corresponds with the current change in climate.

Each of these discoveries and thousands more have contributed to the health and well-being of generations of humans. According to cognitive psychologist Steven Pinker, when the track record of science is objectively evaluated, "we find a substantial record of success—in explanation, in prediction, in providing the basis for successful action and innovation."[5] In *Enlightenment Now*, Pinker explains that through the "awe-inspiring achievements" of science, "we can explain much about the history of the universe, the forces that make us tick, the stuff we're made of, the origins of living things, the machinery of life, including our mental life."[6]

How does science earn this reputation for having such impressive explanatory power? Science is a natural outgrowth of human curiosity. Our ancestors likely had many questions about the natural world, such as whether a plant was safe to eat, why the water was making people sick, or what materials would offer protection from the elements. Science is a systematic and reliable way to collectively pose questions and seek answers about the natural world. Its power comes from scientists' willingness to trust the combined results of many tests, both when these tests support hypotheses and when they do not, an accumulation of scientific consensus. Much of science is not obvious or observable to the individual naked eye. Germs were not understood as the causes of illness until there were microscopes to see them. Although scientists have made significant mistakes, the systematicity and social nature of science allow for self-correction over time. Missteps, wrong assumptions, faulty experiments, and even fraud are eventually uncovered and replaced with ideas that better describe the natural world.

Philosopher of science Lee McIntyre explains how the true value of science comes from adopting a *scientific attitude*, which he describes as an openness to seek new evidence and a willingness to change one's mind in light of evidence.[7] Naomi Oreskes, historian of science, explains in *Why Trust Science?* that while empirical evidence is a cornerstone of science, "it is insufficient for establishing trust in science."[8] The reason the public should place trust in science comes not from individual scientists' contributions, from its mythic "scientific method," or even from specific evidence (as that can change) but rather from the fact that science is a collective enterprise, a social activity. Scientists are fallible and imperfect humans, and there is no single method that leads to some objective truth. Rather, trust comes from "the social character of science and the role it plays in vetting claims."[9] It is through the collective efforts of the social enterprise that scientific consensus is reached.

Examples of the social system that affords science its true value include peer review of scientific articles and grants and governmental organizations such as the National Academy of Sciences. As new evidence comes to light, more accurate views replace those that are flawed. An example of a collective enterprise is the Intergovernmental Panel on Climate Change (IPCC), which derives its conclusions through the work of teams of scientists from diverse backgrounds, multiple countries, and different disciplines, weighing evidence collectively.[10] Through this collaborative effort, the IPCC manages to vet and make public the current, best available evidence on climate change.

If science is such a remarkable achievement and the evidence it has amassed is so compelling, then the roots of resistance beg explanation. Climate scientist Michael Mann cautions that "there is a weakness in the scientific system that can be exploited. The weakness is in public understanding of science, which turns out to be crucial for translating science into public policy."[11] Public misunderstanding of science, coupled with the allure of skepticism, conspiracy theories, or just contrarian points of view, creates a toxic psychological brew. Environmental psychologist Per Espen Stoknes has argued, "We need to look closely at the demand side for doubt—the inner reasons why disbelief is attractive. How does denialism—with very few facts, lots of grand rhetoric, and very little scientific brainpower—continue its dark victory?"[12] Stoknes called for "psychological-level answers" to explain the rising levels of doubt, denial, and resistance prominent today. We intend to do just that in the chapters ahead.

Why Science Is Not Infallible

In this book we call for better-quality and more inclusive science education, science funding, and consideration of scientific consensus in personal and policy decision-making. However, we do not believe science is a panacea for all problems. Nor do we think science is perfect or infallible; it would be foolish to suggest so. Scientists are human beings who can and do make mistakes. Their human characteristics often include brilliance, creativity, imagination, and concern for the health and welfare of fellow human beings and planet Earth. However, individual scientists can also be subject to jealousy, dishonesty, greed, selfishness, and biases that can adversely affect their work.

This is why individual scientists and individual scientific studies should not be given the same level of trust as the collective work of a scientific community amassed over time through rigorous evaluation. It is this consensus that should be widely understood and seriously considered in problem-solving and decision-making, both personally and in education and policy spheres. This is one reason why diversifying science is so important. African American women are woefully underrepresented in the medical field, and they are 3 to 4 times more likely to experience problems with childbirth[13] than their White counterparts. Solutions depend on better representation in both medical and scientific fields.

Even as we call for greater appreciation of science, we acknowledge the history of science as an enterprise replete with mistakes, missteps, sexism, and racism. Science has both helped and hurt individuals and groups who understandably may be wary of accepting scientific claims. There are legitimate concerns with the current scientific enterprise. These include the challenge of replicating findings in psychology and the research dissemination system of journal publication that keeps publicly funded research behind an expensive firewall of inaccessibility. Decreases in public funds for research may motivate researchers to work for entities with vested interests, calling the objectivity of their findings into question. Many critics of science have detailed these and other challenges with science, and we appreciate these efforts to hold science and scientists accountable.[14] The historical and current challenges of science make the call for increased understanding of how science works (and in some cases how it goes awry) all the more important.

But the well-documented racist ideas advanced by science are the most egregious wrongs that must be acknowledged.[15] From the past through to the present, science has propagated false taxonomies of people and cultures to justify European Whites' claims of superiority and used flawed theories of mental measurement of intelligence to justify racial inequity in the United States.[16] If science is to eschew racist ideas and be a force for public good, it must represent the public it serves. This can only be addressed by diversifying science through supporting opportunities for women and people of color, especially African Americans, to be represented in all science fields in numbers that reflect the population. Efforts to diversify science have been made but have yet to reach representative levels.[17] These failures contribute to mistrust of science and hold science back from realizing its full potential to creatively and effectively solve problems that help all citizens.

Why Does It Matter if the Public Understands Science?

The scientific revolution brought civilization the benefits of widespread health and well-being, through fighting disease, providing insights into the natural world, and solving technological challenges.[18] Will science continue contributing solutions to pressing problems? In the midst of a global pandemic, the world's citizens are hopeful yet understandably baffled by what they hear and read. The spread of misinformation and disinformation about science, magnified by a divisive political system and media bubbles, is creating skepticism and mistrust. The common phrase "You can't believe everything you read on the internet" has been expanded by some to "You can't believe anything you read anywhere."

Turning away from science will not solve the complex challenges of climate change, the global spread of the novel coronavirus, clean water, food insecurity, alternative energy production, and a cure for diseases such as Parkinson's and Alzheimer's. Scientific research is needed to understand mass extinctions and perplexing issues such as the rise in rates of food allergies and infertility. Science can play a role in helping to address contemporary problems great and small, and this can be enhanced if the public understands and trusts science to support that role. It also can only play such a role if policy makers learn to listen to scientists and utilize data and evidence to make decisions that affect the lives of others.

Furthermore, adopting a *scientific attitude* is helpful in everyday life, not just in the research lab or field. Individuals with a scientific attitude, as McIntyre describes, care about evidence and are willing to change their minds in light of new evidence. Adopting a scientific attitude might be even more important than taking a deep dive into each nuanced debate on today's scientific topics. Although core consensus remains rather stable over time on many topics, scientific knowledge at the edges is always being updated and revised based on new evidence. A scientific attitude supports understanding the science of tomorrow, not just the science of today.

Maintaining a scientific attitude helps individuals solve problems, sharpen analytical skills, and improve their health and well-being. Science is a process that involves systematically questioning what one knows, revising thinking based on new information, discarding outdated modes of thought, overcoming biases, and using evidence to argue against suspicion and prejudice.[19]

Science Denial, Skepticism, Doubt, and Resistance: What Do These Terms Mean?

The term "science denial" has been used to characterize the thinking of a broad swath of individuals, including those who refuse to accept the scientific consensus on climate change, those who choose not to vaccinate their children, people who view a viral pandemic as a hoax, and the flat-earthers, among others. Deniers are likely to adopt a belief-based attitude,[20] rather than seeking and evaluating the scientific evidence for a claim. Stoknes, in writing about climate deniers, borrows a sociological definition, stating that denial stems from the need to be innocent about a troubling recognition.[21] Fully engaging the idea of climate change is threatening; denial allows an individual to ignore the catastrophic prospects of a deeply troubled planet. Pretending global warming is not happening or that humans have no role in it may lead to fewer sleepless nights, but the implications of placing people in public office and policy positions who take such a stance has far-reaching and potentially irreversible implications for the world. Pronouncing the novel coronavirus pandemic a hoax or dismissing the effects of COVID-19 as no more problematic than the flu has led to the loss of many lives, especially when those erroneous beliefs led to policies that failed to protect citizens.

But are any of these individuals denying science completely or just cherry-picking what claims they might not want to accept? Most individuals are likely to accept antibiotics when needed and to accept that their plane is unlikely to fall from the sky (even if they can't adequately explain the mechanisms of flight), while they may fight fluoridation or refuse vaccinations for their children. McIntyre terms this process "cafeteria denial, as few deny science altogether."[22] Yet there has been a concern in recent years that increasing numbers of individuals are dismissing science in cases where they are not comfortable with the findings or find them contrary to their interests (e.g., elected officials ignoring the projected rise in COVID-19 cases and deaths that would follow reopening of businesses in spring and summer 2020, in order to improve the economy).

Science denial is not, however "science skepticism," a crucial element of science.[23] Skepticism is at the heart of the scientific process, pressing researchers to question deeply, to consider alternative explanations, to actively scrutinize others' work through peer review, and to revise and update theories when the data provide a preponderance of evidence. We all need to be skeptical about claims from questionable sources, whether they advocate using crystals to heal cancer or diets based on eating a single food or injecting bleach to cure COVID-19. This "functional skepticism" is quite distinct, however, from "dysfunctional skepticism," a motivated rejection of science when individuals are only skeptical when the science doesn't support their prior beliefs.[24]

"Science doubt," however, is a serious concern, often used to describe a questioning of science manufactured by vested interests. In *Merchants of Doubt*, Oreskes and Conway describe the history of corporate-funded campaigns to keep people questioning scientific facts long after they have been resolved.[25] This includes the carcinogenic effects of tobacco, the troubling impact of acid rain, the effects of DDT on the ecosystem and individuals, and climate change.

The History of Science Denial, Doubt, and Skepticism

Resistance to science is not new. Yet the current context may differ from any time in the past, given the availability of information and misinformation on the internet and the deliberate attempts to both falsify evidence and willfully disregard it, for corporate or political gain. This current situation resides within a long history of resistance to and denial of scientific knowledge.

Over the last several hundred years, many bold new ideas that upended common understandings took time to take root and gain acceptance. New scientific discoveries often challenge the status quo, and the public may be slow to accept the findings. Sometimes there are vested interests that challenge the new ideas, based not on the data but on the threat to belief systems. Consider Galileo's plight in 1610 when he proposed that Earth orbits the sun, confirming Copernicus's claims that Earth was not the center of the universe. Galileo's observations through a telescope led to conclusions that were condemned by the Roman Catholic Church as contrary to the scriptures, and he was indicted in 1633, forced to recant his ideas, and lived out his life in house arrest.[26] Such conflicts between scientific findings and religious beliefs have played out in other times, perhaps most notably over Charles Darwin's theory of evolution by natural selection, proposed in the *Origin of Species* in 1859. Resistance to evolution was also fueled by essentialist thinking, the idea that natural entities are constant and invariant, as well as by the idea that nature was goal-directed.[27] Darwin's ideas about the role of chance were unacceptable to those who believe that "everything in nature was controlled by necessity."[28]

Scientists, too, have sometimes been slow to ratify consensus on issues that now seem clearly substantiated.[29] Many were notably resistant to Darwin's theories for several decades. The idea of continental drift, proposed in 1912 by geophysicist Alfred Wegener, only gained acceptance 50 years later, when plate tectonics was proposed.[30] The theory that germs caused infection and disease was long resisted, a disregard for the simple finding that when doctors washed their hands, contagion plummeted. Yet many discoveries, such as the structure of DNA, proposed in 1954 by James Watson and Francis Crick, as well as their colleague Rosalind Franklin, garnered quick scientific support. Notably, such a finding does not require a change in attitudes or behavior, nor is it threatening to identity and beliefs.

Science Doubt and Denial in the Modern Era

A cogent example of fostering science denial in the modern era involves the link between smoking and cancer, a connection that had become evident to tobacco companies based on their own research. Beginning in the 1950s, these companies suppressed the research, denied the findings, and then engaged experts to testify on the lack of certainty. Sowing doubt and ambiguity where none existed, they masterminded a campaign of disinformation that went on for decades.[31] As described by Oreskes and Conway in *Merchants of Doubt*,[32] other corporations took note. Admiring the success of the tobacco companies and their ability to delay acceptance of responsibility (and therefore payments to victims) for decades—while continuing to reap massive profits—several major corporations and their consultants took this strategy as their blueprint. Monsanto, producer of multiple pesticides, sought to discredit and defame Rachel Carson, author of *Silent Spring*, a book that not only exposed the effects of DDT but also has been credited with launching the environmental movement in the United States.

Industry's refutation of its role in acid rain (blaming it on natural causes such as volcanos)[33] and DuPont's dismissal of the role that chlorofluorocarbons played in ozone depletion both set the stage for the denial of climate change. In all such cases, the scientific conclusions were resoundingly supported, but actions to mitigate negative effects of commercial products such as tobacco, aerosol sprays, and industrial pollution were deemed costly. Corporations found it easier to sow doubt in the minds of the public and the policy makers. Deniers with advanced degrees can even be purchased,[34,35] lending a thin veneer of credibility to claims about uncertainty and providing talking heads for debates designed to present the illusion of "both sides" of nonexistent controversies. The ethical implications of fostering confusion about settled scientific findings for financial gain are vast. It is no secret who benefits—and who loses.

No other issue involving science denial has more potential for long-term environmental catastrophe than climate change, and that particular denial has been carefully orchestrated. Feeding the psychological bent of the public to place confidence in certainty, oil companies suppressed their own internal research and engaged in a massive effort to foster doubt. As early as 1977, Exxon officials were briefed by one of their senior scientists about the effects on global climate from burning fossil fuels. In 1982, the company's scientists reported further that there were potentially catastrophic effects that might not be reversible.[36] It was not until 2018 that documents emerged showing that Shell, too, had similar knowledge. As environmentalist Bill McKibben notes, the outcome could have been dramatically different had the companies publicly acknowledged what they knew privately and taken the "early inside track on building the energy economy of the future."[37]

What happened instead is that only a month after NASA scientist James Hansen testified in Congress in 1988 and made global warming a public issue, Exxon's public affairs office advocated that the company emphasize the uncertainty of the science of climate change. The coalition of oil companies that formed the Global Climate Coalition used similar strategies to the earlier campaigns regarding tobacco's harm. They even hired the same public relations consultants who manufactured doubt about such ill effects and who orchestrated the attack on Rachel Carson in order to sell more pesticides.[38] In the decades since, the effort to foster uncertainty has become a part of political campaigns bolstered by oil money, and it has been effective. Although the number of Americans who accept that climate change is occurring has risen, the percent who know that scientists agree about the causes is still quite small. A 2018 survey by Yale University found that while 73% of Americans responded that they believe global warming is happening, only 20% of those surveyed understood the high degree of acceptance by scientists.[39] Keeping the illusion of a controversy alive, a conscious and deliberate endeavor, has put money in the pockets of oil companies—and stalled action that could have made a difference in the future of the planet. These practices have continued in multiple areas as environmental regulations have been rolled back. The need to understand science denial, doubt, and resistance is paramount. How do we each keep ourselves from falling for such schemes?

Science Denial and Doubt Today

As the climate crisis illustrates, we are now living at a time when well-substantiated factual claims based on evidentiary science that have been documented, published, and supported through expert consensus can be doubted by large swaths of the public. Moreover, as the novel coronavirus traveled rapidly across the globe and epidemiologists projected its spread and the likely deaths that would result, the United States responded slowly, and political leadership at various levels failed to heed the science and fostered doubt. Even evolution still has resistance, and numerous attempts have been made to teach creationism or "intelligent design" alongside evolution, falsely equating religion and science as two sides of a story. Vaccination hesitancy flourishes, a serious concern at a time that herd immunity is needed to slow and eventually halt a pandemic. Many people find themselves following spurious health claims. Why does it seem worse and more widespread now?

One obvious factor is, of course, the internet and the preponderance of information available to anyone with a smart phone or access to a computer. Our skills for verifying and validating the vast amount of information we each encounter have not kept pace, while the sophistication of those who wish to portray fiction as fact has increased. Social media can amplify our existing beliefs, and people tend to create echo chambers in their media use, hearing more of what they already believe. The corporate "merchants of doubt" have multiplied, with well-funded attempts to willfully ignore, reject, and undermine scientific findings that aren't in their financial interest. Denial has become politicized, and what has been called a "post-truth" era[40] reflects a devaluing of truth. The idea of an "inconvenient truth" (as Al Gore alluded to climate change) is now widespread, seeming to apply to anything a politician would prefer that citizens not believe. In addition, science educators may neglect teaching the underpinnings of science that could help students and future citizens keep alert to attempts to undermine accepted scientific consensus.

What this requires is awareness and vigilance on the part of the public, a willingness to challenge sloppy thinking, spurious claims, the rejection of evidence. We each need to better understand how our own minds and the minds of others work in ways that make it possible to be susceptible to science denial, doubt, and resistance. In a post-truth age of manipulation, we can each be better prepared by becoming aware of cognitive pitfalls, allowing us to practice resistance to anti-science biases.

Why Deny? Psychological Explanations for Science Denial, Doubt, Resistance, and Confusion

In the current era, it is more important than ever to be aware, vigilant, and alert to what seems persuasive and acceptable on the surface. Our goal in this book is to draw broadly on research from psychological science to provide explanations for science denial and doubt.[41] We also strive to help individuals examine faulty thinking and biases (their own, and that of others) as well as to become more effective in discussions or in the presentation of information, whether as colleagues, friends, parents, teachers, or science communicators.

In Chapter 2, "How Do We Make Sense of Science Claims Online?," we address the ways in which the challenges of public understanding of science have been amplified in a digital society. We describe how most people are ill-equipped to engage in the critical reflection that would help them evaluate conflicting claims and biased information or to recognize organizational and corporate agendas that may lead to distorted or one-sided presentations. In this chapter, we review our own research and that of others to describe these growing concerns, introducing several of the key explanatory constructs that will be addressed in more detail in other chapters, and provide examples of how they are relevant to internet searching.

In Chapter 3, "What Role Can Science Education Play?," we explore the state of science knowledge in the United States and the role of science education in preparing a public to effectively evaluate scientific claims. We describe how the current effort to change educational standards has developed to strengthen the US science curriculum and teaching practices and provide examples from our own research on standards-motivated science pedagogy, in both classrooms and informal learning settings such as zoos and museums. We show how instruction from kindergarten to higher education can be used to help students think critically about science topics, weigh evidence against competing theories, and question sources. We also explain the limitations to increasing citizens' science knowledge as the solution to public understanding of science.

In Section II, we systematically describe the basic problems in the understanding and acceptance of science from a psychological perspective. We translate the research on the central psychological issues that foster science doubt and resistance and that challenge the understanding of science. These issues are critical for educators, media specialists, and policy makers to understand if they want to improve science literacy and for individuals to know in order to improve their own ability to clearly interpret the information they encounter.

In Chapter 4, "How Do Cognitive Biases Influence Reasoning?," we give an overview of what social and cognitive psychology can teach us about typical mental pitfalls and how to guard against them. Although individuals may imagine themselves as rational actors, careful and considered in their thinking and capable of sound and reliable judgments, everyone tends to engage in automatic, reflexive thinking rather than in effortful, critical thinking. We describe what it means to be "scienceblind,"[42] when a scientific perspective conflicts with one's own intuitive ideas, and why it is worth turning on the reflective, deliberative mind. We explore how each of us is capable of misjudging the depth of our own understanding on such topics as climate change and how that can be tempered. Numerous cognitive biases also inhibit or override reflective, critical thought, such as "confirmation bias," when individuals seek, interpret, or recall information that aligns with preexisting beliefs. It requires vigilance to guard against these biases, to stay open to new perspectives, and to evaluate information that challenges what you might think you know—or want to believe is true. Learning to do such effortful work and becoming aware of one's own biases and those of others becomes paramount when interpreting complex scientific topics.

In Chapter 5, "How Do Individuals Think About Knowledge and Knowing?," we address the ways in which individuals, in their everyday encounters with new information, conflicting ideas, and claims made by others, decide who and what to believe. Whether anyone is fully aware of it or not, we all hold beliefs about what knowledge is and how we think we know. Can I trust what scientists tell me? What's the best source of information? Do I understand what I am reading, or can I just accept that it is true because of where I read it or who said it? These are questions that involve thinking and reasoning about knowledge, or what psychologists call "epistemic cognition."

We explain how public misunderstanding of scientific claims can be linked to misconceptions about the scientific enterprise itself. Knowing about science involves understanding the premises of science and the practices of science, including the key characteristics of systematic inquiry.

In Chapter 6, "What Motivates People to Question Science?," we explain how individual positions on scientific topics may reflect less familiarity with the topic than core beliefs and worldviews, what psychologists call a "motivated view of science." Even when individuals attempt to be rational and make decisions justified with evidence, they can be biased in what information they attend to and how much credence they give it, as well as the strategies used to assess that information. In addition, humans are social creatures, and people tend to identify with others when they share something in common; thus, views on science may be formed through the lens of social groups. When individuals crowdsource their views on science by polling their social group, rather than evaluating science claims on their own merit, they do not always make scientifically sound decisions.

In Chapter 7, "How Do Emotions and Attitudes Influence Science Understanding?," we challenge the notion of science as a cold, dispassionate enterprise and show how learning about science involves the full range of emotions. Consider the overwhelmingly negative reaction of many members of the public when Pluto was demoted to dwarf planetary status. Emotions are linked to attitudes toward science. When individuals read about topics such as genetically modified foods and stem cell research, they may experience fear or anxiety, which can lead to a negative attitude toward the consumption of food that has been genetically modified or the use of stem cells in medical treatments. However, positive emotions and attitudes such as awe and inspiration can promote better understanding of science. We review the research showing that emotions and attitudes are deeply intertwined in thinking and reasoning about science, as they are with all human experiences.

In Chapter 8, "What Can We Do About Science Denial, Doubt, and Resistance?," we summarize our suggestions for improving public understanding and acceptance of scientific knowledge. Our recommendations are based on the findings from the prior chapters and both summarize and elaborate upon the suggestions in the conclusions of those chapters. For individuals, we address how to think critically about science, how to become aware of cognitive biases, and how to better evaluate scientific information. For educators, we draw on our combined decades of research on teaching critical thinking and reasoning to make practical recommendations that teachers in K–12 and higher education classrooms can adopt to develop science-savvy students, capable of evaluating evidence and making informed decisions. Drawing on a broad body of research on science communication, we provide suggestions for clearer presentation in the media. We also focus on what policy makers can do in the areas of education and science, helping them recognize that science education necessarily includes attention not only to content but also to understanding the nature of science, how it is conducted, and the basic principles of the discipline. Such a focus is essential to combat science denial, doubt, and misunderstanding. We recommend more rigorous teacher preparation standards and encourage education policy makers at all levels to consider what teachers and students need to know about the value and limits of science.

Who Should Read This Book?

If you have chosen to read this book, it is unlikely you are a hard-core science denier.

But maybe you are more likely to trust like-minded peers for answers to questions about scientific topics such as whether to vaccinate your children or to get vaccinated yourself. Perhaps you are troubled by a neighbor who thinks schools should teach "both sides of the evolution controversy" and want to know how to engage in a productive conversation. Or you may have a friend or a relative (or a representative in Congress) whose views on science are so different from yours that you want to try to understand their thinking.

Maybe you find yourself questioning whether the United States should invest in nuclear power to reduce CO_2 emissions, you are curious about the effectiveness of drinking collagen-infused water, or you wonder if it is really necessary to get that flu shot. You may want to understand science and evaluate claims about issues both big and small, global and personal, but you are not sure where to begin. Perhaps you want to know more about the doubts you have developed and want to better understand the workings of your own mind—and that of others. You might be someone who wants better protection against vested interests that try to steer your thinking away from what scientific evidence supports.

We think *educators* who teach about science will find this volume useful. Elementary school teachers may worry about their lack of specific training in science and want to better prepare students to understand science. High school and university science teachers may be frustrated that teaching science content does not always have the expected impact on how students think about scientific issues relevant to their own lives. We want to help educators at all levels appreciate where science doubt and resistance originate and to think about how to teach for deeper understanding and appreciation of the nature of science. We want to support educators in helping students evaluate scientific claims by asking and answering their own questions when they can. We want students to be empowered to evaluate the scientific consensus and trustworthiness of scientific claims when they cannot directly evaluate the evidence themselves. We want to help educators foster the next generation of scientists who will contribute to solving the climate crisis or innovating our energy sector or developing new treatments for medical illnesses. Most importantly, we want to help educators foster a *scientific attitude* in all of their students.

Science *communicators* (journalists, science writers, and scientists) can benefit from reading this volume as well. Science journalists must communicate regularly about controversial science topics while being both fair and persuasive. They are on the front lines of educating the general public about complex scientific issues that inform how people view policies on such topics as mandatory vaccination, tobacco vaping, or needle exchanges, and how they vote on potential issues in their region or community, such as GMO labeling or banning fracking, or how they address those issues with their voting representatives. We believe a better understanding of the psychological issues in this volume might help science journalists enhance their communication efforts.

Scientists increasingly want to learn how to communicate their work to the general public. They strive to reach those who might be skeptical; however, they may not be fully aware of the psychological barriers to effective science communication. Scientists who assume they can do their work and that others will share it with the general public run the risk that their work is either ignored, misunderstood, or misused. Scientists must be an integral part of the efforts to communicate effectively and enrich the public's understanding of science, and this volume could aid them in their efforts.

Policy makers can use the research discussed in these pages to understand points of potential resistance to productive innovation as well as appreciate legitimate concerns about the consequences of science and technology, such as whether we can geoengineer our way out of the climate crisis. Policy makers who improve their understanding and appreciation of science are in a better position to bolster their support for scientific funding and policy initiatives and to make sound policy based on sound science. They may also learn how to better influence individuals to take active steps to improve their health (by quitting smoking, getting vaccinated, etc.). Our suggestions are especially applicable to the "movable middle," those individuals who may be hesitant to take such steps but can be persuaded to do so.

We, like you, perhaps, have often been confused by resistance to science. Too much of our public discourse about science is framed as a political issue, limiting how much science can inform policy and holding back progress, innovation, and greater health and prosperity. Science, when properly conducted and honestly evaluated, brings tremendous benefits to individuals and society. Science can be limited, however, by how much citizens support the enterprise. Science is still our best hope for confronting climate change, protecting our environment, and fighting disease—all issues that must be addressed with their differential impact on low-income people and people of color. This book can serve as a resource for individuals, educators, communicators, scientists, policy makers, and anyone who has pondered the public's understanding of science and has wondered why science denial exists and want to know what to do about it.

Notes

1. Naomi Oreskes, *Why Trust Science?* (Princeton, NJ: Princeton University Press, 2019), 56.
2. Jennifer Marlon, Peter Howe, Matto Mildenberger, Anthony Leiserowitz, and Xinran Wang, "Yale Climate Opinion Maps 2018," Yale Program on Climate Change Communication, August 7, 2018, https://climatecommunication.yale.edu/visualizations-data/ycom-us-2018/?est=happening&type=value&geo=county.
3. Brianne Suldovsky, "In Science Communication, Why Does the Idea of the Public Deficit Always Return? Exploring Key Influences," *Public Understanding of Science* 25 (2016).
4. Patrick Sturgis and Nick Allum, "Science in Society: Re-evaluating the Deficit Model of Public Attitudes," *Public Understanding of Science* 13 (2004).
5. Oreskes, *Why Trust Science?*
6. Steven Pinker, *Enlightenment Now: The Case for Reason, Science, Humanism, and Progress* (New York: Penguin, 2018), 385–86.

7. Lee McIntyre, *The Scientific Attitude: Defending Science from Denial, Fraud, and Pseudoscience* (Cambridge, MA: MIT Press, 2019).

8. Oreskes, *Why Trust Science?*, 57.

9. Oreskes, *Why Trust Science?*, 57.

10. Oreskes, *Why Trust Science?*.

11. Michael E. Mann and Tom Toles, *The Madhouse Effect: How Climate Change Denial Is Threatening Our Planet, Destroying Our Politics, and Driving Us Crazy* (New York: Columbia University Press, 2016), 3.

12. Per Espen Stoknes, *What We Think About When We Try Not To Think About Global Warming: Toward a New Psychology of Climate Action* (White River Junction, VT: Chelsea Green Publishing, 2015), 26.

13. Elizabeth A. Howell, "Reducing Disparities in Severe Maternal Morbidity and Mortality," *Clinical Obstetrics and Gynecology* 61, no. 2 (2018).

14. Cornelia Dean, *Making Sense of Science: Separating Substance from Spin* (Cambridge, MA: Harvard University Press, 2017).

15. John P. Jackson and Nadine M. Weidman, *Race, Racism, and Science: Social Impact and Interaction* (Santa Barbara, CA: ABC-CLIO, 2004).

16. Stephen Jay Gould, *The Mismeasure of Man* (New York: WW Norton, 1981).

17. Julie R. Posselt, *Equity in Science: Representation, Culture, and the Dynamics of Change in Graduate Education* (Stanford, CA: Stanford University Press, 2020).

18. Pinker, *Enlightenment Now*.

19. McIntyre, *The Scientific Attitude*, 202.

20. Stoknes, *What We Think About*.

21. Stoknes, *What We Think About*.

22. Kara Holsopple, "A 'Philosopher of Science' on Climate Change Deniers: 'People Can Actually Change Their Mind Based on Facts,'" September 19, 2019, https://stateimpact.npr.org/pennsylvania/2019/09/19/a-philosopher-of-science-on-climate-change-deniers-people-can-actually-change-their-mind-based-on-facts/.

23. Philipp Schmid and Cornelia Betsch, "Effective Strategies for Rebutting Science Denialism in Public Discussions," *Nature: Human Behavior* 1 (2019).

24. Schmid and Betsch, "Effective Strategies for Rebutting Science Denialism."

25. Naomi Oreskes and Erik M. Conway, *Merchants of Doubt: How a Handful of Scientists Obscured the Truth on Issues from Tobacco Smoke to Global Warming* (New York: Bloomsbury Press, 2010b).

26. Maurice A. Finocchiaro, *Defending Copernicus and Galileo: Critical Reasoning in the Two Affairs* (Dordrecht, The Netherlands: Springer, 2010).

27. Ernst Mayr, "The Ideological Resistance to Darwin's Theory of Natural Selection," *Proceedings of the American Philosophical Society* 135 (1991).

28. Mayr, "The Ideological Resistance to Darwin's Theory," 133.

29. Barnard Barber, "Resistance by Scientists to Scientific Discovery," *Science* 134 (1961).

30. Mayr, "The Ideological Resistance to Darwin's Theory."

31. David Michaels, *Doubt Is Their Product: How Industry's Assault on Science Threatens Your Health* (New York: Oxford University Press, 2008).

32. Oreskes and Conway, *Merchants of Doubt*.
33. Naomi Oreskes and Erik M. Conway, "Global Warming Deniers and Their Proven Strategy of Doubt," *Yale Environment 360*, June 10, 2010a, https://e360.yale.edu/features/global_warming_deniers_and_their_proven_strategy_of_doubt.
34. Mann and Toles, *The Madhouse Effect*.
35. Oreskes and Conway, *Merchants of Doubt*.
36. Bill McKibben, *Falter: Has the Human Game Begun to Play Itself Out?* (New York: Henry Holt and Company, 2019).
37. McKibben, *Falter*, 75.
38. McKibben, *Falter*.
39. Anthony Leiserowitz, Edward Maibach, Seth Rosenthal, John Kotcher, Matthew Ballew, Matthew Goldberg, and Abel Gustafson, *Climate Change in the American Mind* (New Haven, CT: Yale Program on Climate Change Communication, 2018), https://climatecommunication.yale.edu/wp-content/uploads/2019/01/Climate-Change-American-Mind-December-2018.pdf.
40. Lee McIntyre, *Post-Truth* (Cambridge, MA: MIT Press, 2018).
41. We do not address conspiracy theory ideation or focus on political or religious influences on science denial as these issues are addressed extensively elsewhere.
42. Andrew Shtulman, *Scienceblind: Why Our Intuitive Theories About the World Are So Often Wrong* (New York: Basic Books, 2017).

References

Barber, Barnard. "Resistance by Scientists to Scientific Discovery." *Science* 134 (1961): 596–602.

Dean, Cornelia. *Making Sense of Science: Separating Substance from Spin*. Cambridge, MA: Harvard University Press, 2017.

Finocchiaro, Maurice A. *Defending Copernicus and Galileo: Critical Reasoning in the Two Affairs*. Boston Studies in the Philosophy of Science 280. Dordrecht, The Netherlands: Springer, 2010.

Gould, Stephen Jay. *The Mismeasure of Man*. New York: WW Norton, 1981.

Holsopple, Kara. "A 'Philosopher of Science' on Climate Change Deniers: 'People Can Actually Change Their Mind Based on Facts.'" September 19, 2019, https://stateimpact.npr.org/pennsylvania/2019/09/19/a-philosopher-of-science-on-climate-change-deniers-people-can-actually-change-their-mind-based-on-facts/.

Howell, Elizabeth A. "Reducing Disparities in Severe Maternal Morbidity and Mortality." *Clinical Obstetrics and Gynecology* 61, no. 2 (2018): 387–99.

Jackson, John P., and Nadine M. Weidman. *Race, Racism, and Science: Social Impact and Interaction*. Santa Barbara, CA: ABC-CLIO, 2004.

Leiserowitz, Anthony, Edward Maibach, Seth Rosenthal, John Kotcher, Matthew Ballew, Matthew Goldberg, and Abel Gustafson. *Climate Change in the American Mind*. New Haven, CT: Yale Program on Climate Change Communication, 2018. https://climatecommunication.yale.edu/wp-content/uploads/2019/01/Climate-Change-American-Mind-December-2018.pdf.

Mann, Michael E., and Tom Toles. *The Madhouse Effect: How Climate Change Denial Is Threatening Our Planet, Destroying Our Politics, and Driving Us Crazy.* New York: Columbia University Press, 2016.

Marlon, Jennifer, Peter Howe, Matto Mildenberger, Anthony Leiserowitz, and Xinran Wang, "Yale Climate Opinion Maps 2018." Yale Program on Climate Change Communication, August 7, 2018. https://climatecommunication.yale.edu/visualizations-data/ycom-us-2018/?est=happening&type=value&geo=county.

Mayr, Ernst. "The Ideological Resistance to Darwin's Theory of Natural Selection." *Proceedings of the American Philosophical Society* 135 (1991): 125–39.

McIntyre, Lee. *Post-Truth.* Cambridge, MA: MIT Press, 2018.

McIntyre, Lee. *The Scientific Attitude: Defending Science from Denial, Fraud, and Pseudoscience.* Cambridge, MA: MIT Press, 2019.

McKibben, Bill. *Falter: Has the Human Game Begun to Play Itself Out?* New York: Henry Holt and Company, 2019.

Michaels, David. *Doubt Is Their Product: How Industry's Assault on Science Threatens Your Health.* New York: Oxford University Press, 2008.

Oreskes, Naomi. *Why Trust Science?* Princeton, NJ: Princeton University Press, 2019.

Oreskes, Naomi, and Erik M. Conway. "Global Warming Deniers and Their Proven Strategy of Doubt." *Yale Environment 360*, June 10, 2010a. https://e360.yale.edu/features/global_warming_deniers_and_their_proven_strategy_of_doubt.

Oreskes, Naomi, and Erik M. Conway. *Merchants of Doubt: How a Handful of Scientists Obscured the Truth on Issues from Tobacco Smoke to Global Warming.* New York: Bloomsbury Press, 2010b.

Pinker, Steven. *Enlightenment Now: The Case for Reason, Science, Humanism, and Progress.* New York: Penguin, 2018.

Posselt, Julie R. *Equity in Science: Representation, Culture, and the Dynamics of Change in Graduate Education.* Stanford, CA: Stanford University Press, 2020.

Schmid, Philipp, and Cornelia Betsch. "Effective Strategies for Rebutting Science Denialism in Public Discussions." *Nature: Human Behavior* 3, no. 9 (2019): 931–39.

Shtulman, Andrew. *Scienceblind: Why Our Intuitive Theories About the World Are So Often Wrong.* New York: Basic Books, 2017.

Stoknes, Per Espen. *What We Think About When We Try Not to Think About Global Warming: Toward a New Psychology of Climate Action.* White River Junction, VT: Chelsea Green Publishing, 2015.

Sturgis, Patrick, and Nick Allum. "Science in Society: Re-Evaluating the Deficit Model of Public Attitudes." *Public Understanding of Science* 13 (2004): 55–74.

Suldovsky, Brianne. "In Science Communication, Why Does the Idea of the Public Deficit Always Return? Exploring Key Influences." *Public Understanding of Science* 25 (2016): 415–26.

2

How Do We Make Sense of Science Claims Online?

As a new mother, Hannah is often overwhelmed by the decisions she needs to make and by the fragility of her newborn son and the responsibility she has to protect him. As he nears the 2-month mark, she knows her pediatrician will expect her to start the regular regimen of vaccinations recommended for kids. The list he gave her at their last appointment shows that the Centers for Disease Control and Prevention recommend 29 doses of 9 vaccines, all before the age of 6—plus flu shots. In her moms' meeting last week, a group she joined when she wanted advice on breastfeeding, she found comfort in knowing she wasn't alone in trying to decide if this was a good idea or not. Some of the moms have chosen to slow down the process, vaccinate selectively, or not vaccinate at all. The group has a Facebook page, and it's easy to ask for information about any aspect of parenting and receive crowdsourced answers quickly. Like Hannah, her new friends are college-educated and eager to make informed, knowledgeable choices. They send her links to what they have read online and encourage her to think carefully before she takes such a big step, and she decides to do more research herself.

Hannah googles "Should I vaccinate my children?" and gets a page of responses in conflict with one another. "Five important reasons to vaccinate your child" is from something called vaccines.gov, followed by a doctor's testimony, "Dr. Kurt: Why I will never choose to vaccinate my child." She lands on a site called ProCon.org and finds that it's the "leading source for pros and cons of controversial issues." Curious, she looks at the "about" page and finds they are an organization that supports critical thinking and education. Their page on vaccinations gives 10 pros and 9 cons and a cogent summary at the top. Proponents say vaccination is safe, has prevented diseases and saved millions of lives, and adverse reactions are rare. Opponents say that children's immune systems can deal with most infections naturally and that some vaccine ingredients can cause side effects, including seizures, paralysis, and death. Moreover, according to this website, opponents say numerous studies prove that vaccines may trigger autism, attention-deficit hyperactivity disorder, and diabetes.

Science Denial. Gale M. Sinatra and Barbara K. Hofer, Oxford University Press. © Oxford University Press 2021.
DOI: 10.1093/oso/9780190944681.003.0002

The jury must be out on this one, she concludes, after reading back and forth between the two columns and seeing the balance between the sides. She follows up some of the sources cited, including the first one on the con list, the National Center for Vaccination Information, which sounds solid, and their slogan appeals to her: "Your Health. Your Family. Your Choice." Their information seems alarming and further persuades her that she needs to consider this parental option carefully. When she jumps to the second source of information listed, VaxTruth.org, she begins to question why the experts think they are so sure. As the website says, they seem to want us to believe them just because they are the experts and they say so—and Hannah has always been taught to question authority, think critically, and make her own decisions.

Had Hannah found her way to the sites supported by scientific experts and the medical field, she would have been resoundingly reassured that vaccinating her child was the right thing to do, for her family and for her community. The World Health Organization, the Centers for Disease Control and Prevention, and the American Academy of Pediatrics sites all explain with certitude what her doctor had told her about the safety, reliability, value, and importance of vaccinations. What went wrong?

Hannah is not a science denier, although her skepticism about what science can offer has increased in this process. A well-educated, thoughtful person, she has been pulled into the rabbit hole of the internet, having difficulty evaluating what she reads online, wondering how her doctor could be so sure in the face of such persuasive testimony against vaccinating, unclear how some of the friends she most respects can be confident about their alternative choices regarding vaccinations. How does she decide? Where does she place her trust? How does she make sense of her online searching? How does she interpret media attempts at "balance"? How can she enhance her own digital and scientific literacy to navigate the landmines of biased information, algorithmic manipulation, distorted rankings, and false equivalencies and learn to assess expertise effectively? It can seem so very difficult in this era to know what to believe.

Seeking Information in a Digital World

Hannah is not alone in her means of seeking information and her difficulty evaluating it, as you may be likely to recall any number of similarly confusing searches in your own life. In 2019 the Pew Research Center reported that 90% of Americans were online,[1] 81% owned smartphones,[2] and 72% used social media.[3] The internet is a common source of information for nearly any topic, including science. Adolescents, in particular, have reported that their most important sources of knowledge about science come from online resources.[4]

In a 2020 study of digital connectivity worldwide, 77% of the population of 34 countries, on average, were online,[5] a number that continues to grow, as does the number of Google searches—3.8 million *every minute* in a 2019 count. With the proliferation of smartphone personal assistants such as Apple's Siri and devices such as the Amazon Echo and Google Home perched in kitchens and living rooms, seeking an answer to any question simply requires asking it aloud, as even small children are learning, and a source of information is omnipresent on smartphones everywhere. The range of these inquiries goes from the mundane (*Will it rain today? How do I roast a chicken?*) to the complex, many that draw on scientific knowledge: *Is it OK to eat genetically modified organisms? Should I buy an electric car? How do I get more sleep? What happened to the bees that used to pollinate my apple trees? Does my child need meds for hyperactivity? If I take the oxycodone my doctor has prescribed after surgery, will I become addicted? Should my local school be teaching intelligent design alongside evolutionary theory? How much screen time is ok for my kids? What is the cause of climate change—and is it real?* In 2020 many searches were likely to be about whether to use masks during the pandemic, if social distancing was effective, how the novel coronavirus is transmitted, and how long it survives in the air and on surfaces. The need to know became palpably critical for many.

A previously unimaginable range of information is now readily and instantly available, yet that feeling of being overwhelmed by it seems to be passing for many as individuals adapt to a world where information is always at one's fingertips or by simple voice commands. Many people can no longer imagine life otherwise and have grown to appreciate much of what instant accessibility offers. When people are curious about something, when they argue with a friend or colleague, when they want to be amused, when they want to know what's going on in the world, when they need information to make a decision about a complex issue, they turn to the internet. Moreover, individuals face not only a vast array of information but also "misinformation," false content that may be shared by someone who does not realize it as such, and "disinformation," intentionally false information spread for profit or political gain.[6] How do they know what to believe and what is truthful?

How do they weigh this information against the knowledge and experience of friends, family, and professionals?

A Bounded Understanding of Science

An accurate understanding of scientific conclusions is essential for many of the decisions that need to be made, as individuals and as citizens. Yet most people have what research psychologists call a "bounded understanding" of science.[7] They are unlikely to seek or have access to primary articles in peer-reviewed science journals, written for communication within the scientific community (as Barbara's research shows[8]), or to understand them fully if they do, no matter how well-educated they might be in other areas, as they are generally not written for public consumption. Except when deeply committed to developing knowledge in an area (e.g., an environmentalist who seeks to master an understanding of carbon's effect in the atmosphere, a doctor who wants to understand the epidemiology of how a new virus spreads), most individuals face limitations that restrict what they read and what they grasp intellectually. This bounded understanding creates a reliance on those who communicate about scientific findings in more accessible media than in academic journals. As a result, most people must assess scientific claims without a full understanding of the scientific theories, evidence, and findings at the heart of those claims. Cognitive and educational psychologists have investigated how they go about this process, as part of a growing body of research on the public understanding of science.

Evaluating Science Information Online

The vast number of responses when googling any topic requires individuals to choose what to attend to, read/scan, accept, and then act upon, if the search is motivated by a need for a course of action. At the most basic level, we each have to decide what is relevant to our interests and what to ignore, amid a minefield of information. All this happens at lightning speed, often without conscious awareness, what Nobel Prize–winning psychologist Daniel Kahneman[9] has called "System 1," a rapid, instinctive way of responding. The slower, more deliberate and analytical approach of "System 2" thinking, as applied to internet searching, is unlikely to be evoked in most situations. When you want a quick recommendation for a Thai restaurant in a new town or want to settle an argument with a friend about a baseball stat, you are unlikely to spend a lot of time deliberating whether the source of information is the most trustworthy. Yet when individuals activate their System 1 thinking to assess the reliability of more meaningful information or try to evaluate competing claims from multiple sites, they are likely to use a set of their own familiar shortcuts as guides (what are known as "heuristics"). These are often ones that are not mindfully developed and not particularly helpful (e.g., deciding that if it's on the first page of hits, it must be the most accurate and useful). If you further employ your analytical mind, you may assess relevance, evaluate the source, and consider trustworthiness. You might examine whether the claims are supported with evidence and the strength and support for that evidence.

Each person brings to this task particular beliefs about science, often expecting a level of certainty about scientific claims that runs counter to the very basis of scientific findings and explanations as open to revision when warranted by new empirical studies. This expectation can leave some people poorly equipped to adjudicate which claims are well substantiated (e.g., evolutionary theory), perhaps questioning the theory itself when new knowledge elaborates an understanding of evolution (e.g., that modern birds are descended from ancient dinosaurs). Many people were baffled when the Centers for Disease Control and Prevention did not initially advocate for the wearing of masks to prevent contagion of the novel coronavirus in early 2020 and then began to encourage it strongly when more research was available.[10] Without a solid foundation in the tenets of science, people may be distrustful of scientists' embrace of uncertainty and their willingness to revise when new evidence strongly supports an alternative. Understanding the assumptions of the discipline is often not sufficiently a part of science education, but it needs to be.

When individuals seek information from the web, they may make a set of additional judgments: assessing the websites themselves, as well as the authors cited and the claims made, a multifaceted, multilayered process often not in conscious awareness. Trust in both the medium of communication and the source has been described as endangered.[11] Yet many individuals accept what they read, uncritically believing that the top "hits" are likely to be the best, most accurate and useful.[12] When individuals do assess a source on the web, they are likely to consider three components as part of their trustworthiness judgments, according to researchers: *expertise*, *benevolence*, and *integrity*.[13] Basically, this means considering whether the expert is genuinely knowledgeable about the topic and has the credentials that would assuage our doubts, that they are operating without bias for the good of society, and that they are using professionally accepted processes to ascertain the information they are providing. Research from Gale and her colleagues also shows that we should be judging the *plausibility* of claims made, assessing the relative, potential truthfulness. For example, in the case of climate change, middle school students who learned how scientists weigh connections between evidence and scientific ideas were more likely to see the plausibility of how humans are altering earth's environment.[14]

Consumers of online scientific information also need to learn to distinguish between the scientific and nonscientific aspects of an issue.[15] For example, whether foods that are genetically modified can be safely eaten is a different issue from whether the development of genetically modified seeds will lead to monopolies by agribusiness and diminish the role of small farmers. Yet such issues are often confounded in the minds of the public. Many of the scientific issues that befuddle consumers are ones that are embedded in complex social, political, economic, and ethical contexts. Such complexity makes it particularly important to understand not only the knowledge individuals bring to their understanding but their underlying beliefs and attitudes and what counts as expertise and why.

The Valuing of Expertise—and Its Decline

If much of science is too complex for non-experts to fully comprehend, then they are deeply dependent on experts to convey scientific information. The experts relied upon may be scientists themselves or journalists with the training to effectively interpret scientific knowledge or educators whose credentials are trusted. Or instead of seeking the knowledge of experts, people may rely on friends who have made up their minds on an issue or commentators whose opinions on other issues reflect their own or like-minded individuals on Facebook. How do individuals weigh the value of expertise, and how do they decide when it is relevant—and how do they determine who counts as an expert? How much do people trust experts? Do they value scientific consensus?

Recent claims abound that not only has the value of expertise declined but there is a "campaign against established knowledge," as the subtitle of a recent book, *The Death of Expertise*, suggests.[16] Author Tom Nichols argues that ignorance has become a virtue and that society is facing the death of the very *ideal* of expertise, fueled in part by Google, Wikipedia, and the low level of public engagement by experts. What concerns him is not indifference to established knowledge but the rise in hostility toward expert knowledge and the inherent dangers of this stance for a democracy. "The death of expertise is not just a rejection of existing knowledge. It is fundamentally a rejection of science."[17] As noted in the opening vignette of this chapter, some anti-vaccination websites challenge the authority of medical experts in order to make their case, directly appealing to the sensibilities Nichols deplores.

Variation in the value of scientific and medical expertise became particularly evident when the global pandemic spread rapidly. Challenges to such expertise appear to have fueled confusion about effective behavioral norms (e.g., mask wearing, social distancing) as well as about policies that can protect citizens (e.g., when to shut down non-essential businesses and when to reopen). Throughout the early months of the pandemic in 2020, political leaders at various levels demonstrated whether they were using expertise and data to guide their decisions or not, often with tragic consequences.[18]

In spite of the reported overall decline in the value of expertise, US citizens report a relatively high degree of confidence in scientists. This stands in marked contrast to their confidence in the news media, political officials, and business leaders. Of those surveyed by the Pew Research Center in 2016, 84% stated a fair or high degree of confidence in medical scientists and 76% in scientists.[19] Confidence rose with each increase in respondents' educational level. An early 2020 survey by Pew affirmed generally high levels of trust, but showed a large difference by political ideology, with 62% of those who were left-leaning agreeing that they trust scientists a lot to do what is right but only 20% of those who were right-leaning. In spite of a general level of trust by the public, there is large variation between scientists and the public on key scientific issues, with an 18% gap in whether parents should be required to vaccinate their children, a 37% gap on whether climate change is caused by humans, and a 33% difference regarding human evolution over time, a highly substantiated scientific theory. The 2016 report concluded, "Such discrepancies do not happen by accident. In most cases, there are determined lobbies working to undermine public understanding of science: from anti-vaccine campaigners, to creationists, to climate-change deniers." So, while overall trust may appear high, the degree to which individuals think that particular issues have been resolved is often misunderstood, and such doubts may be manufactured and amplified online.

There are also reasons why science is not fully trusted and why healthy skepticism and critical thinking are essential. In spite of professional standards, claims of objectivity, and the peer review process, the conduct of science is also imperfect and can be biased. All experts are not the same, nor do they submit their work to the same scrutiny. Knowing the source of funding can be particularly important in evaluating scientific claims. For example, the Harvard researchers who made claims in the late 1960s about the problems with dietary fat, steering the nation away from perceiving sugar as a culprit in health problems, were funded in part by the sugar industry, through the Sugar Research Foundation. The authors did not disclose their funding source to the *New England Journal of Medicine*, where their influential article appeared, shaping a generation of changes in eating patterns that appears to have fostered higher use of sugar, now widely implicated as a source of the rise in obesity and diabetes.[20] Stories such as this one fuel suspicion—but also lead to further safeguards in the scientific process. Funding disclosures, although not required five decades ago, have since been mandated. This does not mean that the soda, candy, tobacco, and pharmaceutical industries, for example, no longer fund research but only that disclosure is necessary in publication of results. What's critical is that individuals know to look for such funding sources and that policy makers become wary of basing public policy on limited findings.

Moreover, individuals need to learn to assess expertise and ferret out bias. Educators need to help students throughout their education to examine author credentials and training, professional experience, affiliations, and other markers of expertise and find a means for continued education for adults on these topics. Individuals need to learn how to examine the site where the material is posted, reading the "about" section, examining funding sources, seeking out whether there is a specific agenda of the authors, and triangulating the search to find out more about the sponsoring organization and what authority and support it has. Try searching the web for information about "fracking," for example, the process of blasting shale layers far below the surface of the earth to extract natural gas, a controversial scientific topic in the news. If you simply want to know more about what it is, you are quickly led to a site called "Energy in Depth." The home page, called "Just the Facts," begins with asserting and assuming the safety of fracking; the "about" page lets you know that the site was launched by the Independent Petroleum Association of America. Finding unbiased scientific research on the safety concerns about fracking requires considerably more work, as does reviewing these various claims.

Expertise Versus Experience and Personal Testimony

Online expertise is not the only source of information individuals consult. The personal testimony of those with similar experience can be compelling and persuasive, as social psychologists have shown—and it is abundantly available not only among friends and peers but online. Consider Samantha, who is 16 and worried about whether she can get herpes from oral sex. As with every other question she has had about sex and personal health issues, Samantha sometimes talks to her friends and nearly always googles the topic. She knows she will find people online with more experience than she and her friends have, and they will probably have figured out the answers. She likes reading about others' experiences and how they have worked things out for themselves, and she likes the sense of community this gives her. She sometimes contributes back, answering others' questions. Or consider Brad, also 16, who drinks heavily at weekend parties and is concerned about whether he might be an alcoholic. That scares him, given what he has seen happen to his uncle, who is in rehab. Is this something he could have inherited, and if so, is it still ok to drink? He goes online to find out. He knows there are other guys out there who must have faced this problem. Scientists don't have all the answers, he has heard his dad say. He would rather hear from people with real experience.

Like most adolescents with personal questions, Samantha and Brad turn to online resources for help and may find it easier to seek the testimony of their peers' postings than to raise concerns with their parents or doctors or teachers. Adults, too, flock to forums where those who have faced similar issues can describe their own experiences. Often, this can provide help and comfort beyond what doctors have time to offer or provide a form of crowdsourced personal knowledge. If you need treatment for potential skin cancer, you might find it helpful to hear from those who have already endured it and could then know more about what to expect. If you're worried about whether your son's frequent tantrums are normal, you can go beyond the child development books and sites to find out what other parents have observed and how they coped, maybe asking friends on Facebook for advice as they always seem happy to weigh in.

Humans have a preference for information that is vivid and personal. In their research on human inference, psychologists Richard Nisbett and Lee Ross showed how direct experience, case histories, and anecdotes from others are more likely to affect our thinking than abstract information or sta-tistical data.[21] Training to override these cognitive tendencies is paramount. Otherwise, we are at the mercy of the compelling story and the testimony of others or likely to think our own experience is broadly generalizable. One of the more challenging aspects of teaching psychology courses, for example, is helping undergraduates learn to transcend their own immediate expe-rience or their knowledge of the experience of another and to learn to see patterns in data as better evidence. "But my sister had an eating disorder and she recovered quickly" does not decrease the known severity of the disorder or diminish the difficulty of treatment.

Seeing What We Want to See and Googling What We Want to Believe

We all take cognitive shortcuts, and this is particularly evident when seeking information online. Think of any recent Google search you have done and where you stopped the process. How far did you go in your searching and why? Was it when your biases and beliefs were confirmed, or did you find competing information and then figure out how to as-sess it, open to a new resolution of your understanding? As philosopher Michael Lynch notes, we not only see what we want to see, we google what we want to google.[22] If someone has favored politicians who claim that human-induced climate change is a hoax, it is not difficult to find support for that position online—or to enter search terms that lead there. The same was all too true in the early days of the novel coronavirus in the United States, when it was being described as a hoax, in spite of its lethal spread in China and Europe. If you fear that vaccinating your child could result in autism, you might never know that the only article that ever claimed to provide scientific support was denounced as fraudulent and retracted by the journal and that the author, Andrew Wakefield, was stripped of his medical credentials. Unfortunately, it is easy to find the author still lauded as a hero on anti-vaccination websites. If you are told by a fellow parent that Wakefield was persecuted by the medical establishment and want to explore that idea, your search will likely lead to a book entitled *The Truth*, co-written by Wakefield and a television host turned anti-vaccination ac-tivist, Jenny McCarthy. And what if you just want to know if drinking red wine is a good thing or not? Anyone can readily find support for particular desires and beliefs, especially if the search terms are framed in a way that communicates the preferred stance.

"Confirmation bias" is the psychological label for the human tendency to search for and interpret information that supports our existing beliefs and preconceptions. It requires vigilance to guard against it, to stay open to new perspectives, and to evaluate information that challenges what you think you know—or want to believe is true. These limitations are amplified by the architecture and invisible algorithms of internet searching.

Algorithms, Echo Chambers, and Filter Bubbles: The Hidden Structure of Online Searching

The 2016 presidential election made the idea of echo chambers a popular concern, largely in retrospect, as many began to understand that we were not all exposed to the same information about the candidates, daily news, or opinions of others; and by the 2020 election this problem had been further magnified. This is similarly the case in regard to scientific issues, particularly those with political impact, such as climate change. As individuals seek out information, they often go to familiar sources where like-minded individuals gather. If you are reading the *New York Times*, you get a different perspective on current issues than you would from the *Wall Street Journal* or *USA Today*; listening to NPR may give you a view that varies substantially from watching Fox News. Beyond these habits with "conventional media," what we each see on the internet is not what everyone else sees. Algorithms drive that winnowing, often creating what has been termed a "filter bubble."[23] In an age of social media, we each see postings and tweets of friends who share our worldview—and sometimes drop the feeds of those who don't, creating an echo chamber where we basically hear what we already believe. Our positions on issues become reinforced and seldom challenged by alternative perspectives, and we may wonder why others out there are so woefully blind and ignorant of the information that seems visible and obvious to us, no matter what our position might be. Data readily available to anyone about the rise of COVID-19 cases by state during 2020, for example, or the lethal dangers of the virus, may have been less visible to those who got their news from sources eager for economic reopenings.[24]

Individuals need not only to improve their own faculties and strategies for searching but also to become aware of the algorithmic processes that drive search engines and the ways in which what we each see online is shaped, or-ganized, and manipulated. Algorithms are simply a set of rules for solving a problem, such as a recipe or a math equation or computer codes. This basic organizing priniciple of digital life is invisible to users and not something they have access to, even if they want to know. The sequence of results in response to a search is machine-programmed, although how and why are proprietary secrets. What appears in our social media feed can shape beliefs about what is normative, true, and accepted. In an ever-evolving and increas-ingly complex and ubiquitous process, algorithms control what is seen on-line, although many might naively imagine a fairness and objectivity to the process or assume that what appears first in a search is most accurate.

At the most basic level of online searching, Google uses several factors in its algorithm to generate search results and creates a relevancy score from those data, determining how likely one is to see a particular page. The equa-tion involves such factors as how long the page has existed, how many other pages link to it, and the frequency and location of keywords, with search engine "spiders" trolling web content to keep the index growing. Note that the factors for high ranking in a list of results do not include veracity, au-thority, or reliability, none of which can be readily verified by machines, at least at this point in time. This mechanized process, lack of transparency, and vulnerability to manipulation all help explain how a student seeking infor-mation about climate change or a well-educated parent trying to make a de-cision about vaccinations might become bewildered about what is actually true, fueling doubts about science and about what is established knowledge on a topic.

What has concerned many observers is the way in which such practices as those used by Facebook filter what one sees, outside of awareness, that then shape attitudes, beliefs, and behavior. Whether in politics or with scientific information or cat videos, individuals are getting more in their feed of what they have been shown to favor and prefer. If Facebook users have friends who view climate change as a hoax and have indicated they like what they see posted, they will see more of it and can read more of it, and they are not likely to see contradictory information as they are pulled further down the rabbit hole. Accordingly, misinformation multiplies, and users consume more of it over time, reinforcing beliefs. Science denial grows in the process.

Users can even show they like a particular item *before* reading it, which Facebook has determined is a weaker indicator of sentiment than liking it afterward.[25] (Yes, they know these things.) A study of the interactions of

10.1 million US Facebook users found that although algorithms play a role in what we each are exposed to, individuals' choices are significant, and users prefer to read stories that are consistent with their beliefs. As we will explain in other chapters, individuals are typically less rational and objective than they think they are and often seek information that fits their foregone conclusions.

Algorithms play an increasingly powerful role across modern life, well beyond the personalization algorithms of google searching, social media feeds, and purchasing suggestions. Consider any task that involves decision-making and imagine how machine learning might simulate that process. The goal of artificial intelligence engineers is not only to capture what humans might reasonably decide in various scenarios but to improve upon those capabilities to the greatest extent possible. The metrics of the driverless car or diagnostic tools for doctors can arguably do better than a single individual, by harnessing more information and more sources of data. These can be used to create a decision-making rule for a vast set of problems, as well as shorten response times, potentially saving lives in both of these scenarios. But when is human intervention needed? What would happen if a doctor over-relied on such a tool, ignoring the other information available in a doctor–patient interaction?

With algorithms influencing an ever-widening aspect of our lives, the Pew Research Center on Internet, Science, and Technology has suggested that we need not only digital literacy but algorithmic literacy. As one of the experts they surveyed commented, "Unless there is an increased effort to make true information literacy a part of basic education, there will be a class of people who can use algorithms and a class used by algorithms."[26]

What Happens When "Google Knowing" Replaces Understanding?

The ability to google whatever it is that one wants to know may have consequences beyond what any of us have yet to fully grasp. In *The Shallows: What the Internet Is Doing to Our Brains*, Nicholas Carr warns of becoming "pancake people," with no depth of understanding but just a broad collection of shallow bits of information.[27] Philosopher Michael Lynch cautions that an over-reliance on "Google-knowing" may threaten true understanding, which requires learning to reason how information fits together.[28] People may no longer care whether they actually remember a fact, knowing they can look it up again when needed. How does this impede synthesis and creativity, or advanced understanding of a topic, in the absence of a base of genuine knowledge? The broader cognitive effects of being able to glean quick facts digitally and then forget them until they are needed again (and launching another search) are not conducive to genuine understanding of science. Moreover, this pattern of behavior may be attenuating the habits of mind that lead to greater depth of knowing. With a dazzling array of information and all the bells and whistles of modern media, attention spans shrink and individuals learn to skim the surface of information, pluck what is needed or of interest, and then move on, distracted by whatever else is blinking or buzzing—texts, social media feeds, email. Moments later you may have forgotten the very fact you sought.

Facts alone are not knowledge, and knowledge is not understanding. Knowledge requires justification and truth; individuals need to substantiate the information they have gleaned in order to truly know it. Understanding, however, takes this further, advancing to a greater depth of interpretation, allowing for creativity and synthesis to flourish. For example, students who do not know a range of ideas well are not able to assemble them in a new way. They need a command of the material beyond what googling can ever offer. They also need training that will prepare them for this digital world where an abundance of information—some of it biased, conflicting, and erroneous—is always available.

Digital/Media/Information Literacy

Joey visits his grandparents in Cincinnati, and they take him to the Creation Museum, across the bridge in Petersburg, Kentucky. When he makes a report in school, talking about how cool it would have been to ride dinosaurs, his fourth-grade teacher says that dinosaurs and humans did not exist at the same time. (The scientific consensus is that there was a 65-million-year gap between them.) Perplexed by what he saw in a museum and discussed with his grandparents, now contradicted by a teacher, Joey uses the home computer to look it up. He enters "dinosaurs and humans" and looks at the first page of results. His teacher has told him that "dot org" sources are better than "dot com," and the first dot org site is something called the "Institute for Creation Research." Sounds like they would know what they are doing, and the page is well organized and easy to read. Yep, Grandpa was right, he concludes. The site says that available evidence indicates humans and dinosaurs coexisted. On his first page of hits, only one site says it's not true, but that's from RationalWiki, which sounds like Wikipedia, and the teacher doesn't usually want them to use that source.

The next day in school he shows his teacher what he found. She does the search with different terms and finds two sources at the top of her list that support her claim, although Joey notices that both are "dot com" sites. One is from *New Scientist*, on the top 10 dinosaur myths, and the other from Science Views. How does Joey know what to believe? How do our schools help Joey in mastering digital literacy skills over time? How do we help students in the struggle for understanding, especially in situations when perceived authorities in their lives make competing claims about truth?

This is not just a problem for young children. Without training in digital literacy, individuals may be poorly equipped to engage in the critical reflection that would help them discern why, for example, fossil-fuel industrialists present a more skeptical stance on climate change than do insurance-industry representatives. Denial of science is not just an individual matter but is sometimes institutionally organized by those with vested interests, whether this is energy corporations promoting doubt about human causes of climate change or religious organizations disputing evolution. We all need to know how to find information effectively, evaluate what we locate, critically compare what we read, and create an integrated understanding, all aspects of digital and information literacy. As the amount of information available increases exponentially and the means of digital access to it proliferate, training in these skills becomes more important than ever. Gale and her colleague Doug Lombardi have identified steps for how to become better consumers of information online and how to teach others to do the same.[29]

What research has shown is that people often avoid the cognitive work involved in evaluating what they have identified and instead default to judgments based on superficial characteristics. These often include web design—how the website looks, how easy it is to use, as well as how highly ranked the site is in search engine results.[30] These System 1 responses are our default heuristics, and they mean we are less likely to verify information, the more laborious work of System 2. With training, however, it is plausible that new heuristics of evaluation can supplant our surface judgments. Psychologists have also argued that developing digital literacy requires more focus on the cognitive and metacognitive processes involved. This facility with online information also involves self-regulation, including planning, monitoring, and evaluating one's strategies, as well as the awareness to vet and integrate sources.[31]

Digital literacy is far more than learning how to use a computer, although digitalliteracy.gov, the Department of Education website devoted to the topic, focuses mostly on learning technical skills needed in a digital society. Definitional issues abound, and current approaches seem to integrate components of technical skills, competencies at creating content, and skills of evaluating information into a broader category. This need for digital, information, and media literacy is not unacknowledged by educators, although it is often unclear whose role it is. Should instructional technologists offer extra classes for students, or should teachers be prepared to include it in the curriculum? How do educators manage to keep up with the changing technological landscape? Digital literacy requires more than the teaching of a general skill; it also needs embedding and reinforcement within the disciplines. Science educators in particular need to help students learn to navigate the information available to them, on topics not only related to school assignments but more broadly defined. As Barbara's research showed, and research on the problems of knowledge transfer suggests, students are not likely to transfer internet searching skills from the discipline in which they learned them to another; they simply default to their more basic, low-level, general heuristics instead.[32]

Research from Stanford University's Graduate School of Education paints a particularly dismal picture of students' skills at evaluating online information, in a study of nearly 8,000 students across age groups—middle school, high school, and college—and in multiple cities, schools, and universities. Given a series of tasks that required fairly simple evaluation, such as recognition of bias, students were described as "easily duped."[33] The researchers had hoped that middle school students could distinguish an ad from a news story and opinion from news, among other tasks. They hoped that high school students could evaluate evidence and argumentation and would notice that the chart on gun laws was from a political action committee of gun owners. They expected college students to evaluate claims on social media, conduct a search and decide which websites to trust, and see beyond a "dot org" URL to question why only one side of a contentious issue was presented. They were even concerned that the tasks were too easy, but the results indicated just how woefully unprepared most students are, regardless of educational level, to handle the tasks educators expect of them. The good news is that these researchers are now creating curriculum materials for teachers and calling attention to the problem in order to mobilize policy makers. Although it appears students are not becoming skilled in these tasks on their own, they can be taught to be better consumers of information. This attention to digital literacy is critically important in preventing and addressing science denial and doubt, in the school years and throughout the life span.

What Can We Do?

No one can be an expert on all the scientific topics of interest or that require personal decisions, whether these are health issues or environmental concerns or use of the latest technology. Each of us has to decide where to turn for that information and how thoroughly to evaluate it. Sometimes individuals decide their own experience is adequate or the experience of like-minded peers or the opinions of relatives or religious authorities or the judgments of experts whose credentials and training they trust. Hannah, from the opening vignette, trusts her mothers' group, but she also wants to corroborate their views and turns to the internet for further information; yet she may be drawn in by biased sources. In an age of rampant misinformation,[34] knowing what counts as genuine expertise becomes essential, and having the skills to evaluate claims online is increasingly necessary for everyone.

What Can Individuals Do?

Learn to trust the scientific consensus and empirical research over anecdotal accounts. Yes, listen to your own personal experience and that of others, but use it as a starting point and temper it with the wisdom of carefully conducted research. Hannah values her mothers' group and their ideas, but if she reads further and learns about the safety of vaccinations and their merits for both her children and her community, she may realize that anecdotal accounts are limited in their generalizability and represent a limited point of view, sometimes at odds with the present scientific consensus. Erroneous beliefs about vaccines play a significant role in public health regarding COVID-19; some are refusing vaccinations, based on prior beliefs about vaccination safety.[35]

Expand your competence with information literacy, and continue to update your knowledge. Information literacy, defined by the American Library

Association as a set of abilities requiring individuals to "recognize when information is needed and have the ability to locate, evaluate, and use effectively the needed information,"[36] involves skills not only in research but in critical thinking. Such literacy needs lifelong development, given the rapidity of change in the online landscape. If you learned in school, for example, that a "dot org" site was more trustworthy than a "dot com" site, you may be surprised to find that any nonprofit organization can purchase such a site, for any purposes, and often with goals of persuasion. Names of organizations can be carefully crafted to appear non-biased, and the websites can be created to appear professional and reputable. Keep your knowledge current. One way to do this is to go to college library websites that offer advice to their students about searching and evaluating information online. Triangulate that knowledge: go to multiple sites.

Develop skills of evaluating what you encounter online. Evaluate the motives of the article or website (or video or whatever the medium might be). Ask yourself who created the message and why. Are they trying to sell something or promoting a particular point of view? Who are the authors, and what is the organization behind the material? Investigate author expertise through further searching. Click on the "About" heading, and then seek more information about the sponsoring group so that you can assess credibility and potential biases. Look for substantiation of claims, and dig deeply, examining multiple websites for corroboration. Model the behavior of professional fact-checkers, learning to read laterally across sites, in order to verify what you are reading and provide deeper context.[37] Care about how claims are supported and what evidence is provided.

Be aware there are specific steps individuals should take to critically evaluate scientific information found online. In addition to checking facts and evaluating sources, Gale and Doug Lombardi[38] recommend not only examining how well particular sources of evidence support the claim, but also how well the evidence could support an alternative claim. Next, ask yourself, *Is this plausible, and how do I know?* Finally, only share information after you have taken steps to verify it.

Inform yourself about algorithmic literacy. Algorithms are basic formulas for solving problems or completing tasks, such as recipes, math equations, and computer codes.[39] Algorithms determine the outcomes of any internet search, of social media feeds, of ads delivered during searching, and basically of everything conducted online (as well as the decisions made by driverless cars, legal analysis, recruitment selections, financial market trading, etc.). Algorithms are proprietary property, invisible to users, yet many organizations and corporations and individuals learn to game the system, managing to get their own material to move up the ladder of visibility in order to get more viewers, or pay for the privilege. Become aware of this process, and resist thinking that placement in the rankings of any returned search is a proxy for credibility or reliability. Do step outside the bubble that has been created for you, and deliberately seek additional information, contradictory ideas, and reputable findings backed by scientific research. For further information, see reports such as this one by the Pew Research Center for Internet and Technology: "Code-Dependent: Pros and Cons of the Algorithm Age."[40]

What Can Educators Do?

With the internet as a central source of information for students, educators are increasingly pressed to foster information literacy. As noted, this requires both research skills and critical thinking, and it takes ongoing practice with tasks involving careful evaluation of sources. The skills described for individuals are all relevant for students and need to be addressed at every level of education, from elementary school through higher education. Educators can do the following:

Teach information literacy in every class that requires searching for course-related materials online. School libraries may offer workshops for students, but librarians need teachers to reinforce these skills in their assignments. Model your own techniques, and show how you think aloud while searching to evaluate what you find, following the tips from the previous section. Give students practice time in class, monitor their search strategies, and answer their questions.

Recognize that this work of teaching information literacy is never done. As Gale and her colleague Doug Lombardi note, "The rapid pace of technological change, coupled with the misinformation arms race, makes it very difficult to keep up."[41] Continue to polish your own skills, and help keep your students current. Seek professional development opportunities in this area, and if they don't exist, request them.

Teach algorithmic literacy. Help students understand the sequencing of sites that results from any search is the result of a complicated formula that can be manipulated. Help them understand their social media feeds can be providing information that functions as a cocoon of reinforcing ideas and that they need to work to discern truth and to separate fact from fiction.

Teach students how to verify their search results by mimicking the work of fact-checkers. Research show that when students are taught to go beyond reading vertically through a single site but instead learn to read laterally across sites for confirmation, they increase their accuracy in evaluating the credibility of sources.[42] Intervention studies such as this one are proliferating, giving guidance about the type of trainings that are effective.

What Can Science Communicators Do?

Given the "bounded nature" of science, individuals often do not know enough of the science to make an informed decision, nor do they have the time or inclination to read primary articles in scientific journals, even those that are open access (and many are not available without expensive subscriptions). Accordingly, they often read online news sources and websites where they are likely to rely on science journalists and communicators, who can help foster accurate scientific knowledge in a number of ways:

Provide adequate detail about scientific findings to help readers know what conclusions can be drawn. In a medical or health study, for example, did the results come from an experimental clinical trial, or were they correlational findings? Convey meaningful limitations in the study that the authors have stated and others you think seem pertinent. Refer to the sample so that readers can make informed decisions about how far the results might generalize. If a cardiac health study included only White males, for example, say so.

Convey your sources clearly and the impact of the findings. Where was the study published, and why might that matter? Teach the reader about your source of information. How do these findings relate to previous research? What further research is needed to clarify the impact? Who did the research? Foster the valuing of scientific and medical expertise.

Don't oversimplify. German cognitive psychologist Rainer Bromme and his colleagues found that although popularized science reports are usually characterized by simplification, this practice risks "making scientific knowledge seem less complex and easier to evaluate than it actually is."[43] Walk the line between under-reporting and overwhelming the reader. Be aware of what science communication researchers have called "the easiness effect of science popularization."[44]

What Can Policy Makers Do?

Consider the impact of unregulated and unknowable algorithms and how verifiable scientific facts can get buried. This occurs in the "race to the top" that companies and organizations play as they try to get more views of their own positions and try to game the system. Educate yourself about algorithms, engage with the concerns about these topics, create informative conversations that can lead to policies that shape uses and implementation of machine intelligence. For elected representatives, if you haven't done so already, hire a staff person or consultant to keep you current on these issues. Watch how other nations are addressing the same concerns. (Germany has already issued ethics rules to guide the algorithmic choices of autonomous vehicles, such as decision-making when a collision appears imminent.)

Learn from social media companies that have created their own regulatory policies. Examine why and how they have gone about this process. As of May 2019, a search on Twitter for tweets related to vaccines leads directly to a post from the US Department of Health and Human Services, with reliable information and not anti-vax opinions. Pinterest has a policy to prohibit harmful medical misinformation, yet the site was rampant with misleading information and links to anti-vaccination groups. In response, Pinterest made the decision in August 2019 to constrain what users see when they search for information about vaccination safety (and similar searches). Users "can explore reliable results about immunizations from leading public health organizations, including the World Health Organization (WHO), the

Centers for Disease Control and Prevention, the American Academy of Pediatrics, and the WHO-established Vaccine Safety Net, a global network of websites providing reliable vaccine safety information in various languages."[45] Policy makers can take the lead in pressing for accuracy in scientific information across platforms and encouraging this kind of regulation. Imagine what our protagonist Hannah might have encountered differently in her search on vaccination safety were such reforms to take place more broadly.

Value expertise and data in building policy initiatives and guidelines for citizen behavior. Rarely in history has the need for this been more evident than during the global pandemic. Examples abound of decisions being made without regard for the data about viral spread and death tolls, while others take expertise and data into account in determining what to advise.

In the next chapters we explore further the problem of how and why individuals question science and what psychological mechanisms are at work and provide more suggestions for what we can do about this vexing concern.

Notes

1. Pew Research Center, "Internet/Broadband Fact Sheet," June 12, 2019a, https://www.pewresearch.org/internet/fact-sheet/internet-broadband/.
2. Pew Research Center, "Mobile Fact Sheet," June 12, 2019b, https://www.pewresearch.org/internet/fact-sheet/mobile/.
3. Pew Research Center, "Social Media Fact Sheet," June 12, 2019c, https://www.pewresearch.org/internet/fact-sheet/social-media/.
4. Ashley A. Anderson et al., "The 'Nasty Effect': Online Incivility and Risk Perceptions of Emerging Technologies," *Journal of Computer-Mediated Communication* 19, no. 3 (2014).
5. Shannon Schumacher and Nicholas Kent, "8 Charts on Internet Use Around the World as Countries Grapple with COVID-19," Pew Research Center, April 2, 2020, https://www.pewresearch.org/fact-tank/2020/04/02/8-charts-on-internet-use-around-the-world-as-countries-grapple-with-covid-19/.
6. Claire Wardle and Hossein Derakhshan, "Thinking About 'Information Disorder': Formats of Misinformation, Disinformation, and Mal-Information," in *Journalism, 'Fake News' & Disinformation*, ed. Cherilyn Ireton and Julie Posetti (Paris: UNESCO, 2018).

7. Rainer Bromme and Susan R. Goldman, "The Public's Bounded Understanding of Science," *Educational Psychologist* 49, no. 2 (2014).

8. Barbara K. Hofer, "Epistemological Understanding as a Metacognitive Process: Thinking Aloud During Online Searching," *Educational Psychologist* 39 (2004), https://doi.org/10.1207/s15326985Sep3901_5.

9. Daniel Kahneman, *Thinking, Fast and Slow* (New York: Farrar, Straus and Giroux, 2011).

10. Centers for Disease Control and Prevention, "Recommendations Regarding the Use of Cloth Face Coverings, Especially in Areas of Significant Community-Based Transmission," April 3, 2020, https://stacks.cdc.gov/view/cdc/86440.

11. Peter Weingart and Lars Guenther, "Science Communication and the Issue of Trust," *Journal of Science Communication* 15, no. 5 (2016).

12. Hofer, "Epistemological Understanding as a Metacognitive Process."

13. Friederike Hendriks, Dorothe Kienhues, and Rainer Bromme, "Trust in Science and the Science of Trust," in *Trust and Communication in a Digitized World*, ed. Bernd Blöbaum (Cham, Switzerland: Springer, 2016).

14. Doug Lombardi, Gale M. Sinatra, and E. Michael Nussbaum, "Plausibility Reappraisals and Shifts in Middle School Students' Climate Change Conceptions," *Learning and Instruction* 27 (2013).

15. Bromme and Goldman, "The Public's Bounded Understanding of Science."

16. Tom Nichols, *The Death of Expertise: The Campaign Against Established Knowledge and Why It Matters* (New York: Oxford University Press, 2017).

17. Nichols, *The Death of Expertise*, p. 5.

18. Amanda Mull, "Georgia's Experiment in Human Sacrifice," *The Atlantic*, April 29, 2020, https://www.theatlantic.com/health/archive/2020/04/why-georgia-reopening-coronavirus-pandemic/610882/.

19. Brian Kennedy, "Most Americans Trust the Military and Scientists to Act in the Public's Interest," Pew Research Center, October 18, 2016, https://www.pewresearch.org/fact-tank/2016/10/18/most-americans-trust-the-military-and-scientists-to-act-in-the-publics-interest/.

20. Camila Domonoske, "50 Years Ago, Sugar Industry Quietly Paid Scientists to Point Blame at Fat," NPR, September 13, 2016, https://www.npr.org/sections/thetwo-way/2016/09/13/493739074/50-years-ago-sugar-industry-quietly-paid-scientists-to-point-blame-at-fat.

21. Richard Nisbett and Lee Ross, *Human Inference: Strategies and Shortcomings of Social Judgment* (Englewood-Cliffs, NJ: Prentice-Hall, 1980).

22. Michael P. Lynch, *The Internet of Us: Knowing More and Understanding Less in the Age of Big Data* (New York: WW Norton, 2016).

23. Eli Pariser, *The Filter Bubble: How the New Personalized Web Is Changing What We Read and How We Think* (New York: Penguin, 2011).

24. Chris Cillizza, "Here's What You Think of Coronavirus If You Watch Fox News," CNN, April 3, 2020, https://www.cnn.com/2020/04/02/politics/coronavirus-fox-news-poll/index.html.

25. Will Oremus, "Who Controls Your Facebook Feed," *Slate*, January 3 2016), http://www.slate.com/articles/technology/cover_story/2016/01/how_facebook_s_news_feed_algorithm_works.html.

26. Lee Rainie and Janna Anderson, "Code-Dependent: Pros and Cons of the Algorithm Age," Pew Research Center, February 8, 2017, https://www.pewresearch.org/internet/2017/02/08/code-dependent-pros-and-cons-of-the-algorithm-age/.

27. Nicholas G. Carr, *The Shallows: What the Internet Is Doing to Our Brains* (New York: WW Norton, 2010).

28. Lynch, *The Internet of Us*.

29. Sinatra and Lombardi (2020).

30. Nancy A. Cheever and Jeffrey Rokkum, "Internet Credibility and Digital Media Literacy," in *The Wiley Handbook of Psychology, Technology, and Society*, ed. Larry D. Rosen, Nancy Cheever, and L. Mark Carrier (Chichester, England: John Wiley & Sons, 2015).

31. Jeffrey Alan Greene, B. Yu Seung, and Dana Z. Copeland, "Measuring Critical Components of Digital Literacy and Their Relationships with Learning," *Computers & Education* 76 (2014).

32. Hofer, "Epistemological Understanding as a Metacognitive Process."

33. Sam Wineburg et al., "Evaluating Information: The Cornerstone of Civic Online Reasoning," Stanford Digital Repository, 2016, https://purl.stanford.edu/fv751yt5934.

34. Cailin O'Conner and James Owen Weatherall, *The Misinformation Age: How False Beliefs Spread* (New Haven, CT: Yale University Press, 2018).

35. Peter Jamison, "Anti-Vaccination Leaders Seize on Coronavirus to Push Resistance to Inoculation," *Washington Post*, May 5, 2020, https://www.washingtonpost.com/dc-md-va/2020/05/05/anti-vaxxers-wakefield-coronavirus-vaccine/.

36. American Library Association, "Information Literacy," https://literacy.ala.org/information-literacy/.

37. Sam Wineburg and Sarah McGrew, "Lateral Reading: Reading Less and Learning More When Evaluating Digital Information" (working paper 2017-A1, Stanford University, Stanford, CA, October 6, 2017), https://ssrn.com/abstract=3048994.

38. Lombardi and Sinatra (2020).

39. Rainie and Anderson, "Code-Dependent."

40. Rainie and Anderson, "Code-Dependent."

41. Gale M. Sinatra and Doug Lombardi, "Evaluating Sources of Scientific Evidence and Claims in the Post-Truth Era May Require Reappraising Plausibility Judgments," *Educational Psychologist* 55, no. 3 (2020), https://doi.org/10.1080/00461520.2020.1730181.

42. Sarah McGrew et al., "Improving University Students' Web Savvy: An Intervention Study," *British Journal of Educational Psychology* 89, no. 3 (2019), https://doi.org/10.1111/bjep.12279.

43. Rainer Bromme et al., "Effects of Information Comprehensibility and Argument Type on Lay Recipients' Readiness to Defer to Experts When Deciding About Scientific Knowledge Claims" (paper presented at the Annual Meeting of the Cognitive Science Society, 2011).

44. Lisa Scharrer et al., "When Science Becomes Too Easy: Science Popularization Inclines Laypeople to Underrate Their Reliance on Experts," *Public Understanding of Science* 26 (2017), https://doi.org/10.1177/0963662516680311.

45. Ifeoma Ozoma, "Bringing Authoritative Vaccine Results to Pinterest Search," Pinterest, August 28, 2019, https://newsroom.pinterest.com/en/post/bringing-authoritative-vaccine-results-to-pinterest-search.

References

American Library Association. "Information Literacy." https://literacy.ala.org/information-literacy/.

Anderson, Ashley A., Dominique Brossard, Dietram A. Scheufele, Michael A. Xenos, and Peter Ladwig. "The 'Nasty Effect': Online Incivility and Risk Perceptions of Emerging Technologies." *Journal of Computer-Mediated Communication* 19, no. 3 (2014): 373–87.

Bromme, Rainer, and Susan R. Goldman. "The Public's Bounded Understanding of Science." *Educational Psychologist* 49, no. 2 (2014): 59–69.

Bromme, Rainer, Lisa Scharrer, M. Anne Britt, and Marc Stadtler. "Effects of Information Comprehensibility and Argument Type on Lay Recipients' Readiness to Defer to Experts When Deciding About Scientific Knowledge Claims." Paper presented at the Annual Meeting of the Cognitive Science Society, 2011.

Carr, Nicholas G. *The Shallows: What the Internet Is Doing to Our Brains*. New York: WW Norton, 2010.

Centers for Disease Control and Prevention. "Recommendations Regarding the Use of Cloth Face Coverings, Especially in Areas of Significant Community-Based Transmission." April 3, 2020. https://stacks.cdc.gov/view/cdc/86440.

Cheever, Nancy A., and Jeffrey Rokkum. "Internet Credibility and Digital Media Literacy." In *The Wiley Handbook of Psychology, Technology, and Society*, edited by Larry D. Rosen, Nancy Cheever, and L. Mark Carrier, 56–73. Chichester, England: John Wiley & Sons, 2015.

Cillizza, Chris. "Here's What You Think of Coronavirus If You Watch Fox News." CNN, April 3, 2020. https://www.cnn.com/2020/04/02/politics/coronavirus-fox-news-poll/index.html.

Domonoske, Camila. "50 Years Ago, Sugar Industry Quietly Paid Scientists to Point Blame at Fat." NPR, September 13, 2016. https://www.npr.org/sections/thetwo-way/2016/09/13/493739074/50-years-ago-sugar-industry-quietly-paid-scientists-to-point-blame-at-fat.

Greene, Jeffrey Alan, B. Yu Seung, and Dana Z. Copeland. "Measuring Critical Components of Digital Literacy and Their Relationships with Learning." *Computers & Education* 76 (2014): 55–69.

Hendriks, Friederike, Dorothe Kienhues, and Rainer Bromme. "Trust in Science and the Science of Trust." In *Trust and Communication in a Digitized World*, edited by Bernd Blöbaum, 143–59. Cham, Switzerland: Springer, 2016.

Hofer, Barbara K. "Epistemological Understanding as a Metacognitive Process: Thinking Aloud During Online Searching." *Educational Psychologist* 39 (2004): 43–55. https://doi.org/10.1207/s15326985ep3901_5.

Jamison, Peter. "Anti-Vaccination Leaders Seize on Coronavirus to Push Resistance to Inoculation." *Washington Post*, May 5, 2020. https://www.washingtonpost.com/dc-md-va/2020/05/05/anti-vaxxers-wakefield-coronavirus-vaccine/.

Kahneman, Daniel. *Thinking, Fast and Slow*. New York: Farrar, Straus and Giroux, 2011.

Kennedy, Brian. "Most Americans Trust the Military and Scientists to Act in the Public's Interest." Pew Research Center, October 18, 2016. https://www.pewresearch.org/fact-tank/2016/10/18/most-americans-trust-the-military-and-scientists-to-act-in-the-publics-interest/.

Lombardi, Doug, Gale M. Sinatra, and E. Michael Nussbaum. "Plausibility Reappraisals and Shifts in Middle School Students' Climate Change Conceptions." *Learning and Instruction* 27 (2013): 50–62.

Lynch, Michael P. *The Internet of Us: Knowing More and Understanding Less in the Age of Big Data*. New York: WW Norton, 2016.

McGrew, Sarah, Mark Smith, Joel Breakstone, Teresa Ortega, and Samual S. Wineburg. "Improving University Students' Web Savvy: An Intervention Study." *British Journal of Educational Psychology* 89, no. 3 (2019): 485–500. https://doi.org/10.1111/bjep.12279.

Mull, Amanda. "Georgia's Experiment in Human Sacrifice." *The Atlantic*, April 29, 2020. https://www.theatlantic.com/health/archive/2020/04/why-georgia-reopening-coronavirus-pandemic/610882/.

Nichols, Tom. *The Death of Expertise: The Campaign against Established Knowledge and Why It Matters*. New York: Oxford University Press, 2017.

Nisbett, Richard, and Lee Ross. *Human Inference: Strategies and Shortcomings of Social Judgment*. Englewood-Cliffs, NJ: Prentice-Hall, 1980.

O'Conner, Cailin, and James Owen Weatherall. *The Misinformation Age: How False Beliefs Spread*. New Haven, CT: Yale University Press, 2018.

Oremus, Will. "Who Controls Your Facebook Feed." *Slate*, January 3, 2016. http://www.slate.com/articles/technology/cover_story/2016/01/how_facebook_s_news_feed_algorithm_works.html.

Ozoma, Ifeoma. "Bringing Authoritative Vaccine Results to Pinterest Search." Pinterest, August 28, 2019, https://newsroom.pinterest.com/en/post/bringing-authoritative-vaccine-results-to-pinterest-search.

Pariser, Eli. *The Filter Bubble: How the New Personalized Web Is Changing What We Read and How We Think*. New York: Penguin, 2011.

Pew Research Center. "Internet/Broadband Fact Sheet." June 12, 2019a. https://www.pewresearch.org/internet/fact-sheet/internet-broadband/.

Pew Research Center. "Mobile Fact Sheet." June 12, 2019b. https://www.pewresearch.org/internet/fact-sheet/mobile/.

Pew Research Center. "Social Media Fact Sheet." June 12, 2019c. https://www.pewresearch.org/internet/fact-sheet/social-media/.

Rainie, Lee, and Janna Anderson. "Code-Dependent: Pros and Cons of the Algorithm Age." Pew Research Center, February 8, 2017, https://www.pewresearch.org/internet/2017/02/08/code-dependent-pros-and-cons-of-the-algorithm-age/.

Scharrer, Lisa, Yvonne Rupieper, Marc Stadtler, and Rainer Bromme. "When Science Becomes Too Easy: Science Popularization Inclines Laypeople to Underrate Their Reliance on Experts." *Public Understanding of Science* 26 (2017): 1003–18. https://doi.org/10.1177/0963662516680311.

Schumacher, Shannon, and Nicholas Kent. "8 Charts on Internet Use Around the World as Countries Grapple with COVID-19." Pew Research Center, April 2, 2020. https://

www.pewresearch.org/fact-tank/2020/04/02/8-charts-on-internet-use-around-the-world-as-countries-grapple-with-covid-19/.

Sinatra, Gale M., and Doug Lombardi. "Evaluating Sources of Scientific Evidence and Claims in the Post-Truth Era May Require Reappraising Plausibility Judgments." *Educational Psychologist* 55, no. 3 (2020): 120–131. https://doi.org/10.1080/00461520.2020.1730181.

Wardle, Claire, and Hossein Derakhshan. "Thinking About 'Information Disorder': Formats of Misinformation, Disinformation, and Mal-Information." In *Journalism, 'Fake News'& Disinformation*, edited by Cherilyn Ireton and Julie Posetti, 43–54. Paris: UNESCO, 2018.

Weingart, Peter, and Lars Guenther. "Science Communication and the Issue of Trust." *Journal of Science Communication* 15, no. 5 (2016): C01.

Wineburg, Sam, and Sarah McGrew. "Lateral Reading: Reading Less and Learning More When Evaluating Digital Information." Working paper 2017-A1. Stanford University, Stanford, CA, October 6, 2017. https://ssrn.com/abstract=3048994.

Wineburg, Sam, Sarah McGrew, Joel Breakstone, and Teresa Ortega. "Evaluating Information: The Cornerstone of Civic Online Reasoning." Stanford Digital Repository, 2016. https://purl.stanford.edu/fv751yt5934.

3

What Role Can Science Education Play?

When Carmen relocated from Los Angeles to Spokane for her dream job after completing her master's degree, she took her young son to a new dentist. When the dentist recommended fluoride drops, Carmen was confused. "Why do I need to give him drops? Isn't there fluoride in the water?" The dentist explained that even though the American Dental Association (ADA) recommends adding fluoride to the public water supply, the citizens of Spokane had voted against it. Carmen went home with her son's prescription but also with questions. She had heard fluoride was safe, and she thought it was in the water supply everywhere. Apparently, some people are opposed to fluoride, but why? What were the benefits and risks? Was it safe for a child to take fluoride drops? Carmen searched online.

The phrase "Is fluoride safe" returned over 28 million results. One website indicated that fluoride was perfectly safe and was recommended by the ADA. But another stated that it was unnecessary at best and harmful at worst. The more Carmen read, the more confused she became about why this was considered controversial. With a reasonably strong background in science and a sense of which websites provide research-based information, Carmen searched the American Academy of Pediatrics, the ADA, and the Centers for Disease Control and Prevention websites for answers. These websites restated what the dentist had told her: fluoride is safe, it should be in drinking water, and if not, consumers should use drops. Then why had the citizens of her new hometown banned fluoride if medical evidence suggested it was not only safe but beneficial?

Carmen is an educated, concerned parent trying to learn about the potential benefits and risks to her family of water fluoridation. Plenty of information is available on this well-researched topic, and if she takes the time to download articles from reputable medical sites and evaluate the evidence for and against fluoridation, she is likely to make a well-informed decision about what's best for her young son. However, not everyone makes sound decisions when evaluating scientific issues. Even those with a strong knowledge base may fall

Science Denial. Gale M. Sinatra and Barbara K. Hofer, Oxford University Press. © Oxford University Press 2021.
DOI: 10.1093/oso/9780190944681.003.0003

prey to compelling conspiracy theories or may double down on their original conceptions (or misconceptions). It is easy to acquire new misconceptions from reading questionable content on the internet that is hard to distinguish from more reliable sources. Even after evaluating the evidence, taking actions based on scientifically accurate knowledge is not always enhanced by additional information alone. Knowledge is important, but it is not always sufficient to prompt sound scientific judgments and actions. But what kind of knowledge does matter? What kind of science education can help?

What Do Americans Know About Science?

Everyday decision-making, such as whether to vaccinate one's children, finish a course of antibiotics, eat genetically modified foods, or wear a mask to reduce the spread of a virus, is informed by what individuals know (or believe they know) about the scientific consensus on the matter. Understanding (or lack of understanding) of science has a direct day-to-day impact in terms of health, wellness, and safety. When you fail to vaccinate your children, do not finish a course of antibiotics, avoid foods your body needs, or do not maintain social distance during a pandemic, the health and safety of you and your family are put at risk.

What do American citizens know about science? How do we compare in our knowledge to other countries? Is greater science knowledge the key to greater science acceptance? These are questions that have vexed researchers and education policy makers for decades, and they are not as easy to answer as they may seem. The Pew Science Knowledge Quiz, an 11-item multiple-choice test of basic science knowledge, shows that the average score of Americans was 6.7 correct or 61%,[1] a grade of D in a US classroom. The Oxford Scale, which has been repeated periodically since the 1980s, shows that in 2014 Americans were able to answer about 6 out of 9 questions, such as *Does the earth go around the sun or does the sun go around the earth?* This level of performance has remained relatively stable since the first administration.

But there is much more to understanding science than can be revealed by a science "pop quiz" that predominately assesses fact-based knowledge. More critical is understanding how scientists know what they know and how they adjudicate knowledge claims. "More than just basic knowledge of science facts, contemporary definitions of science literacy have expanded to include understanding of scientific processes and practices, familiarity with how science and scientists work, a capacity to weigh and evaluate the products of science, and an ability to engage in civic decisions about the value of science" (p. 1).[2] For example, when asked what does it mean to study something scientifically, only 26% of Americans could give an adequate response. These other aspects of science knowledge are not well assessed by fact-based knowledge quizzes but are likely much more important features of public understanding of science.

How Do Americans Measure Up?

Concerns about scientific literacy in the United States have been persistent for decades. The Soviet Union's launch of Sputnik in 1957 kicked off a competitive space race to close the perceived science and technology gap between the Soviet Union and the United States. Education in science, technology, engineering, and mathematics (the STEM disciplines) was infused with billions of dollars of funding. Just a few years later, President Kennedy promised NASA would put a man on the moon. The infusion of energy and resources into STEM education was a boost to the economy, improved United States standing in the world in technology and science, and likely motivated many students to pursue science education and careers who might not have otherwise.[3]

An examination of international comparative tests, such as the Programme for International Student Assessment (PISA), administered every 3 years to 15-year-olds around the globe, reveals that today US students do appear to lag behind those in other developed nations in scientific knowledge. In 2018 the US performed below about a dozen other countries in science and 36th in math.[4] This level of comparative performance has stayed relatively the same over the years, dating back to the original administration of this assessment in 1967, leading some to question whether efforts to improve test scores were having any impact.[5]

The science knowledge arms race is problematic in several ways. Comparisons to other countries are exceedingly difficult. While the tests themselves may have some utility in particular contexts, the comparison of positions of one country versus another can be misleading. In some top-scoring countries, the population of available PISA test takers is comprised of those students who have advanced beyond the compulsory schooling that stops in the 9th grade (e.g., China), skewing their results upward when compared to nations where most students continue their education beyond age 15.[6] Even actual scores on the tests are often not comparable; it is not as if each country's scores are determined by a simple sum of the number of correct answers, but rather scores are determined through intricate metrics that complicate direct comparison.[7] International trends can also obscure actual growth in US students' knowledge over time as other countries are not stagnant but improve in their science efforts or change their policies regarding who is actually tested, moving their scores up or down. In sum, "the use of country rankings (especially compared over time) is very likely to be misleading about levels of students' achievement."[8]

Even if direct comparisons to other countries are not particularly informative, those concerned about US students' relative science preparation note persistent issues in US science education that may lead to comparatively poor performance internationally. These include a relative lack of emphasis on science in the curriculum compared to other countries, poor preparation of science teachers, underpaid teachers, textbooks that cover a vast number of topics and few in depth, and critiques of how science is taught. For example, the number of high school courses US students take in science can be as low as one.[9] Considering that roughly one-third of Americans do not go on to higher education and that those who do may never take another science course, scientific literacy depends on greater K–12 science preparation. What students learn in these science courses, however, is related to their teachers' preparation in science and science education. In the US, elementary school teachers are not required to have a science background, and most take one, if they take any, science course during college. Most states have no requirements for passing subject matter tests for elementary teacher certification.[10] At the high school level, teachers are required to take subject matter tests in only 38 states, and even in those states the policies are not sufficient to assure that teachers have adequate content knowledge.[11] Even more problematic, the number of teachers asked to teach outside their expertise due to staff shortfalls is alarming. Roughly 30% of physical science teachers in high school did not major in these fields and are not certified to teach these subjects.[12] Beyond teacher preparation, others point to the low pay for teachers relative to other professions requiring a 4-year degree.

Pervasive socioeconomic and educational inequities exist in the United States, such that higher-income students receive more rigorous and higher-quality STEM instruction than lower-income students.[13] For example, high–socioeconomic status (SES) students are 4 times more likely to take calculus (a required course for advanced STEM degrees in college) and 3 times more likely to take advanced science courses in high school compared to their lower-SES peers.[14] Teachers with the least STEM preparation are more likely to end up teaching in low-SES schools and districts, giving low-SES students little chance to improve their STEM preparation.

In part due to the lack of science education they have received themselves and in part due to modest requirements for science education courses in most teacher preparation programs, many K–12 teachers report they are not confident in their own skills for teaching science, especially in ways consistent with current views on how science should be taught.[15] Strengthening K–12 teachers' preparation to teach science, along with heightening science education standards, would very likely improve K–12 students' science knowledge and might help to fill the STEM positions many argue are needed for the growth of the American economy.[16]

Is It Just a Knowledge Deficit?

There are good arguments to improve the quality of science education in the United States, but the notion that public understanding of science can be easily fixed by following the simple recipe of "just add knowledge"[17] reflects a *knowledge deficit view* that has been roundly criticized. The notion that more knowledge is the singular fix has intuitive appeal, but it is overly simplistic. The idea that if we just communicated the right way about scientific knowledge, the public's understanding and acceptance would significantly improve is simply not borne out by the data. Barbara's[18] and Gale's[19] own research, as well as a meta-analysis examining dozens of studies,[20] shows that knowledge has an inconsistent relationship with science attitudes and acceptance in that more knowledge is sometimes linked to greater acceptance and more positive attitudes, and sometimes it is not.[21] Science communicators who may feel a responsibility to close the gap between what scientists and members of the public know have themselves at times fallen prey to the idea that direct transfer of knowledge is possible.[22] But increasingly they have come to realize that heads are not empty vessels waiting to be filled with scientific information. Rather, heads exist on the shoulders of whole persons who have beliefs, emotions, attitudes, motivations, and social and cultural group memberships, which all contribute to framing individuals' understanding of science.

As educators, we are never going to argue against efforts to increase the science knowledge of students and citizens. But we also recognize the limits of this approach. Even as we write this, it is evident that communicating the scientific facts about mask wearing during a pandemic is not just a straight-forward matter of conveying correct information about the mask's effectiveness in guarding against the spread of respiratory droplets. The physics of the matter is not that difficult to convey or understand, and we encourage everyone to watch one of the many video demonstrations showing how aerosol mists float in the air after a sneeze or cough to become better informed. But we also recognize that individuals may be as (if not more) persuaded to wear a mask by seeing their favorite politician or their Facebook friends donning masks as they are by learning more about the physics of respiration.

Not Just More but Different Science Education

Traditional science education has long been criticized for relying too heavily on lessons involving dry textbook passages, demonstrations, or "cookbook" lab experiments. These well-intentioned lessons go back to the late 1800s in the United States[23] and were meant to engage students in scientific methods and the inquiry process but did not always succeed in their efforts. You may remember making a baking soda and vinegar "volcano" explode on your lab bench in chemistry class or using a pH sensor to test hypotheses about acid–base concentrations in a solution. Watching demonstrations and conducting experiments are mainstays of science instruction, but experiments involve more than following step-by-step procedures in a lab book to derive a predetermined answer to a question students may not even care about.[24] Do you know what chemical principle that baking soda volcano you made in elementary school was supposed to illustrate or why you conducted an acid–base titration in your high school chemistry course?

Such approaches were often designed to teach the steps of "the scientific method" many of us remember learning about from elementary school science lessons through high school chemistry class. These steps include formulating questions, generating hypotheses, conducting an experiment, analyzing data, interpreting results—all steps that many scientists do engage in to conduct their research. Science educators, once enamored with teaching "the" scientific method, have more recently come to appreciate that this is "a simplistic account of the process of science."[25] Scientists use a variety of methods, and not every scientific study involves each of the traditional steps completed in a linear fashion.[26] Collecting observational data is often just as important as experiments in fields where not every question can be answered through controlled studies. Consider the field of evolutionary biology where the discovery of a fossil can realign the lineage of an entire species, such as the recent discovery of a fossil that changes our view of how modern humans are related to ape-like ancestors.[27]

A hallmark of science, no matter which methods are used, is a systematic and deliberative process. Certainly, careful scientists strive for systematicity; however, even this view portrays science as a paint-by-numbers endeavor where scientists seek simple answers through following a set of one-size-fits-all procedures. The history of science is replete with famous examples of key scientific insights made by creative guessing, random chance, luck, or even mistakes. Saccharin, the artificial sweetener, was discovered when a researcher failed to wash his hands before dinner, picked up a dinner roll, and discovered a chemical from his lab experiment made it taste sweet.[28] Penicillin was serendipitously derived when a scientist noticed mold growing on a Petri dish had prevented bacterial growth. In other words, science is a human endeavor and, as such, is much more fluid and creative than following a cookbook approach that gets to "Truth" with a capital "T." Research on the development of a coronavirus vaccine illustrates the nature of science as researchers try to address a global pandemic in real time with the eyes of the world riveted. The dire need for preventative vaccines and disease treatment raised concerns in members of the public (and in some science circles) that the traditional steps of vaccine development might be skipped to advance remedies at a rapid pace, which could reduce public confidence in these potential advancements.[29] The transparency of the final stages of the US Food and Drug Administration's vaccine review and approval process, as illustrated by the live broadcast of the Emergency Use Authorization panel meeting, did assuage concern for some skeptics.

Against the backdrop of a new understanding of the nature of science, educators called for students to be actively engaged in the practices of science, such as asking and answering their own questions. The Next Generation Science Standards (NGSS) for K–12 education in the United States were developed to reform science education toward greater engagement of students.[30] The basic framework emphasizes the practices that scientist engage in to do their work, such as asking questions, designing systematic ways to answer those questions, and then sharing their findings. In other words, rather than a focus on teaching *only* science content (such as the periodic table, genetic inheritance, the process of photosynthesis) or doing cookbook demonstrations or labs, instruction inspired by the new standards asks students to engage in the practices of doing actual science.

Since its inception, 20 states had adopted the NGSS standards by 2019.[31] Since scientific practices are not well captured by a textbook or a set of lab experiments, educational researchers have been striving to develop materials and activities that educators can adopt to engage their students in asking and answering their own questions. Uncertainty remains whether K–12 teachers, especially those in elementary school, are prepared to teach science in this more authentic manner, given the limits of current science teacher preparation and the lack of curriculum resources available that meet the standards.

What does NGSS instruction look like in elementary school? Gale and her team, inspired by the new standards, considered that one way to engage students in scientific practices would be to leverage young children's natural interest in play. Partnering with the Mattel Children's Foundation, they worked with elementary school teachers to develop an NGSS-aligned curriculum using Hot Wheels cars and tracks they called "Speedometry."[32] The goal was to have elementary school students learn about concepts such as force and motion, traction, and engineering design while engaging in scientific practices such as asking and answering their own questions. Students in 4th grade designed and conducted their own experiments to see which cars would go faster than others and whether adding different materials to the cars or tracks would speed them up or slow them down. Students shared their findings with peers. The curriculum extends down to kindergarten, with activities designed to promote understanding of concepts such as friction, force, and motion in these young learners.

There are several takeaways the team learned through their work on the Speedometry project. First, if science is taught through engaging hands-on activities where students have choice in deciding the questions they want to ask and answer, they are drawn to it. Students in the Speedometry groups demonstrated greater understanding of concepts such as force and motion and reported that they were more interested in the content than were students in a control group.[33] The toys may have helped get the students interested initially, but research interviews revealed that it wasn't all about the play. Students reported they enjoyed learning about the science content, too.[34] If we want to increase the amount of science instruction that occurs in K–12 education, it will be critical to assure that science learning is active, personalized, relevant, and engaging. We think Speedometry accomplished that, but there are many other examples of active science instruction that teachers can use in their own instruction.[35]

Teachers also welcomed the Speedometry curriculum. They admitted they had not been teaching a lot of science due to the pressure to teach what is most frequently assessed on national standardized tests, literacy and math,[36] and appreciated having these activities to try out in their own classrooms. Some were a bit apprehensive, admitting they often shied away from teaching science because they didn't feel confident enough in their own science knowledge to support their students' learning. Speedometry also provided teachers two things they lacked, a curriculum and materials that were aligned to the new standards and the support needed to implement it.

What does NGSS instruction look like in secondary school? Teachers don't need toys to engage students in scientific reasoning. For example, middle school students are capable of reasoning about evidence just as scientists do. In one study, conducted by Doug Lombardi of the University of Maryland, middle school students were asked to evaluate evidence as to whether human activity was contributing to climate change. Students had to decide whether each piece of evidence supported the scientific view that human activity is the key source of the problem or whether the evidence supported a contrary view popular among climate skeptics (increased solar radiation). These middle schoolers were more than capable of evaluating the evidence. The researchers found that the exercise of thinking like a scientist resulted in greater understanding that human activity is the driving force behind climate change. Students who engaged in this activity not only showed a greater understanding of humans' role in climate change than students who learned the same information in a more traditional form of instruction but actually retained that understanding better than a control group when retested months later. Some adults struggle to find it plausible that human activity can shift global climate patterns.[37] And yet, the students in this study who evaluated the evidence found the scientific model more plausible than the skeptic model.

This evidence-evaluation activity was originally developed by Clark Chinn and his colleagues at Rutgers University to move students away from simplistic step-by-step inquiry projects and toward engaging in the type of scientific practices that the NGSS standards recommend. The students in Chinn and Lombardi's research studies engage in model-based reasoning about other topics, such as evolution and fracking, with equal success.[38] Their work illustrates how engaging students in scientific practices of thinking and reasoning like a scientist promotes more than increased science knowledge; it changes how they think about science itself.

What does NGSS instruction look like outside of schools?   Learning about science also happens in places other than schools and universities. Museums, zoos, and other informal learning spaces are a major source of science learning for the general public.[39] Recently, Gale teamed up with scientists at La Brea Tar Pits and Museum in Los Angeles and multimedia designers at the University of Southern California's Institute for Creative Technologies to design a new augmented reality exhibit.[40] In this exhibit, visitors will see a virtual mammoth become stuck in the asphalt seeps (tar pits) as virtual dire wolves come to prey on the distressed beast. They will also be able to experience, through the augmented reality exhibit, the process scientists use to excavate the remains of the ill-fated mammoth and her predators, which also succumbed to gooey asphalt. They will determine whether the fossil belonged to an extinct saber-toothed cat that no longer roams the streets of Los Angeles or some creature that still makes its home in the Hollywood hills, such as a mountain lion. Preliminary research on a pilot version of the exhibit shows that augmented reality/virtual reality have great potential for science learning in informal educational contexts, which is why many museums are capitalizing on this new means of engaging visitors.[41] These technologies are particularly helpful for visualizing something that is difficult to see (such as microscopic organisms), inaccessible (a deep-sea octopus in its natural habitat), or hard to even imagine (what did Los Angeles look like 50,000 years ago?).

Not Just More but Different Knowledge

Students arrive in classrooms with prior knowledge, which largely supports their learning about science. But often science learners in K–12 and higher education, as well as the general public, have misconceptions that can make learning science more challenging.

Learning something new that you know relatively little about can be easier than having to overcome misconceptions and restructure your thinking, a process psychologists call "conceptual change."[42] Imagine a child thinks

Earth is flat, but her teacher tells her Earth is round. The young learner will likely have a difficult time understanding this new concept of Earth's shape because the surface she walks on every day is flat. What will be needed is a significant change in her understanding of Earth's shape. As she learns this new information she may go through a revision of her knowledge from flat Earth to a round pancake-like shape (a common intermediate step) to finally a view of Earth as round like a sphere.[43]

Researchers who study conceptual change know that when teaching about scientific topics where learners of any age have misconceptions, a good place to start is by upending those ideas. In a series of studies, Gale and her colleagues have been exploring misconceptions students and members of the general public have about genetically modified organisms (GMOs) that are consumed as food. Many genetic modifications in the foods we eat every day come from cross-pollination, which is a natural process plant breeders have used for thousands of years, producing hybrid organisms. In contrast, the term "GMO" generally refers to organisms whose genetic material (DNA) has been modified through the introduction of a gene from a different organism,[44] leading some to apply the derogatory label "Frankenfood."[45] There are several misconceptions people have about the process of genetic modification. For example, some think GMOs are created by injecting hormones into a plant or animal.

In several related studies on GMOs, Gale and her colleagues have employed an approach that has been widely used to confront scientific misconceptions called a "refutation text."[46,47] Refutation texts are designed to point out specific misconceptions that a reader of scientific content is likely to have about the topic and debunk them.[48] These texts are most effective when they employ a three-part structure, first calling attention to the misconception with a statement like "Some people think injecting hormones into a plant or animal produces a genetic modification." The next step is to refute that misconception by saying "This is not correct" or "However, this is not the case." The third step, a critical one, is to provide evidence supporting the correct scientific conception. So, for example, the text could say "This is not correct. Injecting hormones into a plant or animal can increase its growth rate or size. However, injecting hormones does not modify the genetic makeup of the plant or animal."

Gale and her colleagues, as well as many other scholars in psychology and education, have used refutation texts (or a similar refutational style of classroom teaching) to confront scientific misconceptions about vaccination,[49] climate change,[50] GMOs,[51] and a number of other controversial topics. They have also been used effectively to overcome misconceptions about less controversial topics such as why the seasons change.[52] Researchers have used a variety of ways to test the effectiveness of refutation texts, but the general approach is to compare the outcomes on assessment of correct knowledge or misconceptions before and after reading the refutation text compared to the performance of those who read similar information presented in a traditionally structured text. Both texts are often structured like an article online or a textbook passage, with only one directly refuting misconceptions. Overall, the research has tended to show an advantage for refutation texts over expository texts in decreasing misconceptions.[53]

Misconceptions about topics like GMOs are often associated with negative attitudes and emotions. "Frankenfood" makes some individuals conjure up images that are nothing like how most foods are actually modified. However, when these misconceptions are cleared up, not only do individuals have a better understanding of what GMOs are but their negative attitudes and emotions are reduced.[54] Why is this important? There are ballot initiatives popping up in states such as California where individuals have to decide whether to put a GMO label on foods that have been modified. It would be difficult to make a well-reasoned decision about labeling when misconceptions are clouding judgments. Virtually all corn in the US food supply has been modified at some point in its history, and corn is in many food products and used as feed for cows and other animals. This makes labeling a challenging process to get right. There is a multimillion-dollar food industry jumping on the labeling bandwagon; however, labeling foods "non-GMO" or "GMO-free" can be misleading. Companies are rushing to label their food "GMO-free" even when it doesn't make sense. You can find Himalayan salt labeled "GMO-free" despite the fact that salt is a rock and doesn't have genes. This is like labeling an iPad as "GMO-free." Consumers are willing to pay more for a product labeled "GMO-free,"[55] so their decisions should be based on facts, not fear and confusion over what GMOs are and whether they are harmful to our health.

Without a strong science education background, consumers are susceptible to unsubstantiated claims they do not have the knowledge to evaluate. Perhaps nowhere is this more evident than the rash of claims made about potential treatments for COVID-19. Not only were drugs like hydroxychloroquine touted as major breakthroughs in the absence of strong evidence for their use in the treatment for COVID-19 but also a rash of other supposed remedies, such as zinc and vitamin D, were flying off the shelves. Whether it is an unproven drug with potentially harmful side effects or supplements or vitamins that could show some benefit,[56] consumers need to have the skills to evaluate whether treatments have been shown to be both safe and effective.[57]

When Scientifically Accurate Knowledge Is Still Not Enough

Science educators on our research teams and others have successfully confronted individuals' misconceptions about climate change, human evolution,[58] genetically modified foods,[59] vaccinations,[60] and Pluto's demotion[61] to dwarf planetary status. We have also developed instruction in how to think critically about science topics, weigh evidence against competing theories, and question sources. We have engaged students from kindergarten through college and members of the general public in a variety of activities from reading texts that confront misconceptions to designing and conducting their own experiments to watching augmented reality experiences of mammoths getting stuck in tar pits. All of these approaches have met with some successes and, along with many others in the field of science education, show that more and different science instruction can help address the science literacy challenge.

We also need to recognize the limits to increasing knowledge as the solution to public acceptance of scientific issues. Given the complexity of these issues, the rapid pace of new discoveries, and conflicting viewpoints available online, most of us cannot fully understand all the science before we have to make a decision on an issue. Individuals must make decisions such as whether to buy flood insurance for a home that may be closer to the water's edge in 10 years than it is now due to sea level rise without adequate knowledge. Researchers found that after Hurricane Irma hit Cape Verde, individuals expressed more certainty that climate change contributed to hurricanes, had more negative emotions about climate change, and were more willing to pay higher taxes to mitigate climate change effects than before the hurricane.[62] Experiencing a superstorm may shift individuals' thinking, but perspectives on climate change do not necessarily shift when just based on new scientific information.[63]

In a recent study Gale conducted with her students, they presented climate scientist Michael Mann's well-known "hockey stick" graph showing the dramatic rise in CO_2 levels corresponding with the rise of industrialization.[64] This particular hockey stick graph, although similar in form to others seen every day, has been extremely controversial because the graph demonstrated that temperature had risen with the increase in industrialization, implicating fossil fuels as a driver of climate change. The graph created a firestorm as those who feared increased government regulation to curb CO_2 tried to undermine the implications drawn from the graph.[65]

In Gale's study, some online participants answered questions about the CO_2 hockey stick graph, and other participants were asked the *same* questions about the *same* graph relabeled as depicting something other than CO_2 rise. One group saw the graph labeled as though it represented the rise in housing costs over time, while another saw the graph relabeled to purportedly show the rise in autism diagnoses. Participants saw only one of the three graphs, either the original depicting CO_2 rise or one of the "doctored" graphics ostensibly depicting two other unrelated (and fictitious) trends showing the same pattern.

Overall, politically conservative participants were particularly bad at interpreting the graph when it depicted changes in CO_2.[66] Conservatives were less likely to attribute the rise in global temperature to increased CO_2 levels as the graph illustrates than politically more liberal participants. However, conservatives were more likely than liberals to select other causes such as natural variation in global temperatures. Conservatives who saw either of the other two graphs did not have the same difficulty interpreting the trends. In other words, they did not tend to think they were seeing natural variations in housing costs or autism rates.

Factors other than knowledge, such as political affiliation, values, and motivations, may come into play when reading a graph about climate change or evaluating the risks and benefits of fluoridation or deciding whether GMO foods need labeling.

What Can We Do?

Improving science instruction would no doubt help the public navigate challenging science topics they need to understand to make informed decisions.

However, more knowledge is not the main problem nor the simple solution to public understanding of science. Science reform efforts need to address broader science literacy. Science educators, communicators, and policy makers can contribute in multiple ways.

What Can Educators Do?

Educators from K–12 to colleges and universities, as well as educators in informal settings, know that learners bring their own ideas and knowledge with them when learning about science. Sometimes these ideas can be in conflict with the current scientific consensus. Educators, as a first step, should seek to understand what misconceptions students may have about the topic and then to help them learn the correct scientific conception. Respect for individuals' worldviews is laudable, but some ideas that are at odds with science can be harmful (such as the view that vaping nicotine is not harmful to your health, when mounting evidence suggests that it is).[67] Everyone (scientists, students, and citizens) must use the best evidence to decide whether they accept or do not accept scientific explanations. Deciding what actions to take (wearing a mask in public during a pandemic) or not (vaping) works best when informed by evidence. Once you understand what GMOs are, you are in a better position to decide what to eat and how to vote on labeling initiatives. Educators can help individuals understand how to engage in the processes scientists use to evaluate evidence and make informed decisions.

New methods of teaching based on engaging learners of all ages in the practices of science are becoming more widespread, and the evidence is mounting for their effectiveness. We encourage active engagement of learners in scientific practices. These practices should be taught to all students, not just those aspiring to careers in science-related fields. Scientific practices aid decision-making in many areas of life. Should I replace my old appliances with more costly energy-efficient models? Will I get a return on investment if I put solar panels on the roof? Is it safe to use pesticides on my backyard vegetable garden?

What Can Science Communicators Do?

Science communicators are experts at explaining and translating complex concepts to their audiences, and many do so masterfully. They provide a critical public service to society through their efforts. They should also be aware of misconceptions the public may have about any given topic and how to confront misconceptions with direct refutations followed by coherent explanations of the scientific phenomena when the science is clear. Equally important is providing the context for how to understand emerging evidence when the science is not yet settled. Presenting "both sides" of a settled scientific issue can be harmful. "Fair and balanced" is neither fair nor balanced when unsupported views are given equal time to scientific perspectives. Being clear about what areas of science lack consensus is equally important. Too often, writers share news about the "cures" or "treatments" or "diets" that are effective or not effective based on single studies. Stating the limits of science (such as the tentativeness of the findings) should also be clear in science communications.

What Can Policy Makers Do?

Meeting the needs of the ever-growing STEM workforce cannot be done without support for science education at all levels. The return on investment to the economy for every dollar spent on scientific research has been estimated to be close to 30% or more depending on the specific investment.[68] Support for public funding for scientific research has been dwindling, but spending less money on scientific research comes at its own cost. Science funded by corporations does not have the same level of independence and can result in biased outcomes that undermine the public trust in science overall.

Local policy makers are in a position to support science education in both formal (public schools and state universities) and informal (museums, zoos, and community education) spaces. But the benefits to society of a scientifically literate population extend far beyond the economy. As we noted, scientific reasoning skills can be applied to decision-making in many disciplines and in our day-to-day lives.

Conclusions

Misconceptions about science are common, and more knowledge is usually not a quick and easy fix. Students in formal and informal settings learn best about science when they have the opportunity to engage in some of the same practices that scientists use such as asking and answering their own questions in systematic ways that take evidence into account. Other factors such as how individuals think and reason about knowledge and their motivations and emotions can be even more important than knowledge in determining whether science is supported or resisted.

Notes

1. Brian Kennedy and Meg Hefferon, "What American Know About Science," Pew Research Center, March 28, 2019, https://www.pewresearch.org/science/2019/03/28/what-americans-know-about-science/.
2. National Academies of Sciences, Engineering, and Medicine. *Science literacy: Concepts, contexts, and consequences*. Washington, DC: National Academies Press, 2016.
3. John L Rudolph, *How We Teach Science—What's Changed, and Why It Matters* (Cambridge, MA: Harvard University Press, 2019).
4. Jill Barshay, "What 2018 Pisa International Rankings Tell Us About U.S. Schools," The Hechinger Report, December 16, 2019, https://hechingerreport.org/what-2018-pisa-international-rankings-tell-us-about-u-s-schools/.
5. Dana Goldstein, " 'It Just Isn't Working': Pisa Test Scores Cast Doubt on U.S. Education Efforts," *New York Times*, December 5, 2019, https://www.nytimes.com/2019/12/03/us/us-students-international-test-scores.html.
6. T. Loveless, *Brown Center Report on American Education: How Well Are American Students Learning? Part III: A Progress Report on the Common Core* (Washington, DC: Brookings Institution, 2014).
7. Judith Torney-Purta and Jo-Ann Amadeo, "International Large-Scale Assessments: Challenges in Reporting and Potentials for Secondary Analysis," *Research in Comparative and International Education* 8, no. 3 (2013).
8. Torney-Purta and Amadeo, "International Large-Scale Assessments."
9. Jason Koebler, "Many Stem Teachers Don't Hold Certifications: Shortages Force Educators to Teach Subjects Outside of Their Specialty Areas," *U.S. News & World Report*, June 8, 2011, https://www.usnews.com/education/blogs/high-school-notes/2011/06/08/many-stem-teachers-dont-hold-certifications.
10. National Council on Teacher Quality, "Content Knowledge: Elementary Teacher Preparation Policy," May 2019a, https://www.nctq.org/yearbook/national/Content-Knowledge-75.
11. National Council on Teacher Quality, "Secondary Content Knowledge: Secondary Teacher Preparation Policy," April 2019b, https://www.nctq.org/yearbook/national/Secondary-Content-Knowledge-84.
12. Koebler, "Many Stem Teachers Don't Hold Certifications."
13. William H. Schmidt et al., "The Role of Schooling in Perpetuating Educational Inequality: An International Perspective," *Educational Researcher* 44, no. 7 (2015).
14. National Science Foundation, "Science & Engineering Indicators 2018," 2018, https://www.nsf.gov/statistics/2018/nsb20181/

15. Susan Haag and Colleen Megowan, "Next Generation Science Standards: A National Mixed-Methods Study on Teacher Readiness," *School Science and Mathematics* 115, no. 8 (2015).

16. Smithsonian Science Education Center, "The STEM Imperative," 2020, https://ssec.si.edu/stem-imperative.

17. Gale M. Sinatra and Robert W. Danielson, "Adapting Evolution Education to a Warming Climate of Teaching and Learning," in *Evolutionary Perspectives on Child Development and Education* (Cham, Switzerland: Springer, 2016).

18. Barbara K. Hofer, C. F. Lam, and A. DeLisi, "Understanding Evolutionary Theory: The Role of Epistemological Development and Beliefs," in *Epistemology and Science Education: Understanding the Evolution vs. Intelligent Design Controversy*, ed. R. Taylor and M Ferrari (New York: Routledge, 2011).

19. Gale M. Sinatra et al., "Intentions and Beliefs in Students' Understanding and Acceptance of Biological Evolution," *Journal of Research in Science Teaching* 40, no. 5 (2003).

20. Nick Allum et al., "Science Knowledge and Attitudes Across Cultures: A Meta-Analysis," *Public Understanding of Science* 17, no. 1 (2008).

21. Sinatra et al., "Intentions and Beliefs in Students' Understanding."

22. Brianne Suldovsky, "In Science Communication, Why Does the Idea of the Public Deficit Always Return? Exploring Key Influences," *Public Understanding of Science* 25 (2016).

23. Rudolph, *How We Teach Science.*

24. R. Driver and J. Easley, "Pupils and Paradigms: A Review of Literature Related to Concept Development in Adolescent Science Students," *Studies in Science Education* 5 (1978).

25. Rudolph, *How We Teach Science.*

26. Lee McIntyre, *The Scientific Attitude: Defending Science from Denial, Fraud, and Pseudoscience* (Cambridge, MA: MIT Press, 2019).

27. Rob Picheta, " 'Missing Link' in Human History Confirmed After Long Debate," CNN, January 19, 2019, https://www.cnn.com/2019/01/19/health/australopithecus-sediba-human-history-scli-intl/index.html.

28. Kevin Loria, "These 18 Accidental and Unintended Scientific Discoveries Changed the World," *Science Alert*, April 4, 2018, https://www.sciencealert.com/these-eighteen-accidental-scientific-discoveries-changed-the-world.

29. Paul K. Komesaroff, Ian Kerridge, and Lyn Gilbert, "The US Is Fast-Tracking a Coronavirus Vaccine, but Bypassing Safety Standards May Not Be Worth the Cost," *The Conversation*, March 20, 2020, https://theconversation.com/the-us-is-fast-tracking-a-coronavirus-vaccine-but-bypassing-safety-standards-may-not-be-worth-the-cost-134041.

30. NGSS Lead States, *Next Generation Science Standards: For States, by States* (Washington, DC: National Academies Press, 2013).

31. National Science Teaching Association, "About the Next Generation Science Standards," 2014, https://ngss.nsta.org/About.aspx.

32. Speedometry, "Math and Science Curriculum: Stem Lesson Plans & Activities," 2020, http://origin2.hotwheels.mattel.com/en-us/explore/speedometry/index.html.

33. Morgan Polikoff et al., "The Impact of Speedometry on Student Knowledge, Interest, and Emotions," *Journal of Research on Educational Effectiveness* 11, no. 2 (2018).

34. Gale M. Sinatra et al., "Speedometry: A Vehicle for Promoting Interest and Engagement through Integrated Stem Instruction," *Journal of Educational Research* 110, no. 3 (2017).

35. Doug Lombardi, "The Curious Construct of Active Learning," *Psychological Science in the Public Interest* (in press).

36. Ananya M. Matewos et al., "Teacher Learning from Supplementary Curricular Materials: Shifting Instructional Roles," *Teaching and Teacher Education* 83 (2019).

37. Doug Lombardi and Gale M. Sinatra, "College Students' Perceptions About the Plausibility of Human-Induced Climate Change," *Research in Science Education* 42, no. 2 (2012).

38. Doug Lombardi, "Beyond the Controversy: Instructional Scaffolds to Promote," *The Earth Scientist* 32, no. 2 (2016).

39. National Research Council, *Learning Science in Informal Environments: People, Places, and Pursuits* (Washington, DC: National Academies Press, 2009).

40. Diane Krieger, "Breaking Down the Emotional Barriers to Science Learning: Professor Gale Sinatra Brings a Real-World Approach to the Public Understanding of Science," *Rossier Magazine*, 2018, https://rossier.usc.edu/magazine/ss2018/breaking-emotional-barriers-science-learning/.

41. Alana U. Kennedy et al., *Re-Living Paleontology: Using Augmented Reality to Promote Engagement and Learning* (San Francisco, CA: American Psychological Association, 2018).

42. S. Vosniadou, ed. *International Handbook of Conceptual Change* (New York: Routledge, 2008).

43. S. Vosniadou and W. F. Brewer, "Mental Models of the Earth: A Study of Conceptual Change in Childhood," *Cognitive Psychology* 24 (1992).

44. World Health Organization, "Food, Genetically Modified," 2020, https://www.who.int/health-topics/food-genetically-modified/#tab=tab_1.

45. Iina Hellsten, "Focus on Metaphors: The Case of 'Frankenfood' on the Web," *Journal of Computer-Mediated Communication* 8, no. 4 (2003).

46. Benjamin C. Heddy et al., "Modifying Knowledge, Emotions, and Attitudes About Genetically Modified Foods," *Journal of Experimental Education* 85, no. 3 (2016).

47. Ian Thacker et al., "Using Persuasive Refutation Texts to Prompt Attitudinal and Conceptual Change," *Journal of Educational Psychology* 112, no. 6 (2020).

48. Gale M. Sinatra and Suzanne H. Broughton, "Bridging Reading Comprehension and Conceptual Change in Science: The Promise of Refutation Text," *Reading Research Quarterly* 46, no. 4 (2011).

49. Erica D. Kessler, Jason L. G. Braasch, and Carolanne M, Kardash, "Individual Differences in Revising (and Maintaining) Accurate and Inaccurate Beliefs About Childhood Vaccinations," *Discourse Processes* 56, no. 5–6 (2019).

50. Robert W. Danielson, Gale M. Sinatra, and Panayiota Kendeou, "Augmenting the Refutation Text Effect with Analogies and Graphics," *Discourse Processes* 53, no. 5–6 (2016).

51. Heddy et al., "Modifying Knowledge, Emotions, and Attitudes."

52. J. Cordova et al., "Self-Efficacy, Confidence in Prior Knowledge, and Conceptual Change," *Contemporary Educational Psychology* 39 (2014).

53. Sinatra and Broughton, "Bridging Reading Comprehension and Conceptual Change."

54. Heddy et al., "Modifying Knowledge, Emotions, and Attitudes."

55. Christopher C. Bruno and Benjamin L. Campbell, "Students' Willingness to Pay for More Local, Organic, Non-GMO and General Food Options," *Journal of Food Distribution Research* 47, no. 3 (2016).

56. Fiona Mitchell, "Vitamin-D and Covid-19: Do Deficient Risk a Poorer Outcome?," *Lancet Diabetes & Endocrinology* 8, no. 7 (2020).

57. Gale M. Sinatra and Doug Lombardi, "Evaluating Sources of Scientific Evidence and Claims in the Post-Truth Era May Require Reappraising Plausibility Judgments," *Educational Psychologist* 55, no. 3 (2020).

58. Patricia H. Hawley and Stephen D. Short, "The Effects of Evolution Education: Examining Attitudes Towards and Knowledge of Evolution in College Courses," *Evolutionary Psychology* 13, no. 1 (2015).

59. Thacker et al., "Using Persuasive Refutation Texts."

60. Kessler, Braasch, and Kardash, "Individual Differences in Revising."

61. Suzanne H. Broughton, Gale M. Sinatra, and E. Michael Nussbaum, " 'Pluto Has Been a Planet My Whole Life!' Emotions, Attitudes, and Conceptual Change in Elementary Students' Learning About Pluto's Reclassification," *Research in Science Education* 43 (2013).

62. Magnus Bergquist, Andreas Nilsson, and Wesley Schultz, "Experiencing a Severe Weather Event Increases Concerns About Climate Change," *Frontiers in Psychology* 10 (2019).

63. Dan M. Kahan et al., "The Polarizing Impact of Science Literacy and Numeracy on Perceived Climate Change Risks," *Nature Climate Change* 2, no. 10 (2012).

64. Robert W. Danielson et al., "When Strategic Graphical Interpretation Fails: The Influence of Prior Belief and Political Identity" (poster presented to the European Association for Research on Learning and Instruction, Tampere, Finland, August, 2017).

65. Michael E. Mann, *The Hockey Stick and the Climate Wars: Dispatches from the Front Lines* (New York: Columbia University Press, 2013).

66. Danielson et al., "When Strategic Graphical Interpretation Fails."

67. Michael Joseph Blaha, "5 Vaping Facts You Need to Know," Johns Hopkins Medicine, 2020, https://www.hopkinsmedicine.org/health/wellness-and-prevention/5-truths-you-need-to-know-about-vaping.

68. Sheila Campbell and Chad Shirley, "Estimating the Long-Term Effects of Federal R&D Spending: CBO's Current Approach and Research Needs," Congressional Budget Office, June 21, 2019, https://www.cbo.gov/publication/54089.

References

Allum, Nick, Patrick Sturgis, Dimitra Tabourazi, and Ian Brunton-Smith. "Science Knowledge and Attitudes Across Cultures: A Meta-Analysis." *Public Understanding of Science* 17, no. 1 (2008): 35–54.

Barshay, Jill. "What 2018 PISA International Rankings Tell Us About U.S. School." The Hechinger Report, December 16, 2019. https://hechingerreport.org/what-2018-pisa-international-rankings-tell-us-about-u-s-schools/.

Bergquist, Magnus, Andreas Nilsson, and Wesley Schultz. "Experiencing a Severe Weather Event Increases Concerns About Climate Change." *Frontiers in Psychology* 10 (2019): 220.

Blaha, Michael Joseph. "5 Vaping Facts You Need to Know." Johns Hopkins Medicine, 2020. https://www.hopkinsmedicine.org/health/wellness-and-prevention/5-truths-you-need-to-know-about-vaping.

Broughton, Suzanne H., Gale M. Sinatra, and E. Michael Nussbaum. " 'Pluto Has Been a Planet My Whole Life!' Emotions, Attitudes, and Conceptual Change in Elementary Students' Learning About Pluto's Reclassification." *Research in Science Education* 43 (2013): 529–50.

Bruno, Christopher C., and Benjamin L. Campbell. "Students' Willingness to Pay for More Local, Organic, Non-GMO and General Food Options." *Journal of Food Distribution Research* 47, no. 3 (2016): 32–48.

Campbell, Sheila, and Chad Shirley. "Estimating the Long-Term Effects of Federal R&D Spending: CBO's Current Approach and Research Needs." Congressional Budget Office, June 21, 2019. https://www.cbo.gov/publication/54089.

Cordova, J., Gale M. Sinatra, S. H. Broughton, G. Taasoobshirazi, and Doug Lombardi. "Self-Efficacy, Confidence in Prior Knowledge, and Conceptual Change." *Contemporary Educational Psychology* 39 (2014): 164–74.

Danielson, Robert W., Gale M. Sinatra, and Panayiota Kendeou. "Augmenting the Refutation Text Effect with Analogies and Graphics." *Discourse Processes* 53, no. 5–6 (2016): 392–414.

Danielson, Robert W., Gale M. Sinatra, Ian Thacker, and Neil Jacobson. "When Strategic Graphical Interpretation Fails: The Influence of Prior Belief and Political Identity." Poster presented to the European Association for Research on Learning and Instruction, Tampere, Finland, August 2017.

Driver, R., and J. Easley. "Pupils and Paradigms: A Review of Literature Related to Concept Development in Adolescent Science Students." *Studies in Science Education* 5 (1978): 61–84.

Goldstein, Dana. " 'It Just Isn't Working': PISA Test Scores Cast Doubt on U.S. Education Efforts." *New York Times*, December 5, 2019. https://www.nytimes.com/2019/12/03/us/us-students-international-test-scores.html.

Haag, Susan, and Colleen Megowan. "Next Generation Science Standards: A National Mixed-Methods Study on Teacher Readiness." *School Science and Mathematics* 115, no. 8 (2015): 416–26.

Hawley, Patricia H., and Stephen D. Short. "The Effects of Evolution Education: Examining Attitudes Towards and Knowledge of Evolution in College Courses." *Evolutionary Psychology* 13, no. 1 (2015): 67–88.

Heddy, Benjamin C., R. W. Danielson, G. M. Sinatra, and J. Graham. "Modifying Knowledge, Emotions, and Attitudes About Genetically Modified Foods." *Journal of Experimental Education* 85, no. 3 (2017): 513–33.

Hellsten, Iina. "Focus on Metaphors: The Case of 'Frankenfood' on the Web." *Journal of Computer-Mediated Communication* 8, no. 4 (2003): JCMC841.

Hofer, Barbara K., C. F. Lam, and A. DeLisi. "Understanding Evolutionary Theory: The Role of Epistemological Development and Beliefs." In *Epistemology and Science Education: Understanding the Evolution vs. Intelligent Design Controversy*, edited by R. Taylor and M. Ferrari, 95–110. New York: Routledge, 2011.

Kahan, Dan M., Ellen Peters, Maggie Wittlin, Paul Slovic, Lisa Larrimore Ouellette, Donald Braman, and Gregory Mandel. "The Polarizing Impact of Science Literacy and Numeracy on Perceived Climate Change Risks." *Nature Climate Change* 2, no. 10 (2012): 732–35.

Kennedy, Alana U., Neil Jacobson, Ian Thacker, Gale M. Sinatra, X. Lu, J. H. Sohn, David Nelson, E. S. Rosenberg, and Ben D. Nye. *Re-Living Paleontology: Using Augmented Reality to Promote Engagement and Learning.* San Francisco, CA: American Psychological Association, 2018.

Kennedy, Brian, and Meg Hefferon. "What Americans Know About Science." Pew Research Center, March 28, 2019. https://www.pewresearch.org/science/2019/03/28/what-americans-know-about-science/.

Kessler, Erica D., Jason L. G. Braasch, and Carolanne M. Kardash. "Individual Differences in Revising (and Maintaining) Accurate and Inaccurate Beliefs About Childhood Vaccinations." *Discourse Processes* 56, no. 5–6 (2019): 415–28.

Koebler, Jason. "Many Stem Teachers Don't Hold Certifications: Shortages Force Educators to Teach Subjects Outside of Their Specialty Areas." *U.S. News & World Report*, June 8, 2011. https://www.usnews.com/education/blogs/high-school-notes/2011/06/08/many-stem-teachers-dont-hold-certifications.

Komesaroff, Paul K., Ian Kerridge, and Lyn Gilbert. "The US Is Fast-Tracking a Coronavirus Vaccine, but Bypassing Safety Standards May Not Be Worth the Cost." *The Conversation*, March 20, 2020. https://theconversation.com/the-us-is-fast-tracking-a-coronavirus-vaccine-but-bypassing-safety-standards-may-not-be-worth-the-cost-134041.

Krieger, Diane. "Breaking Down the Emotional Barriers to Science Learning: Professor Gale Sinatra Brings a Real-World Approach to the Public Understanding of Science." *Rossier Magazine*, 2018. https://rossier.usc.edu/magazine/ss2018/breaking-emotional-barriers-science-learning/.

Lombardi, Doug. "Beyond the Controversy: Instructional Scaffolds to Promote." *The Earth Scientist* 32, no. 2 (2016): 5–10.

Lombardi, Doug, Shipley, T. F., Astronomy Team (Bailey, J. M, Bretones, P. S., Prather, E. E.), Biology Team (Ballen, C. J., Knight, J. K., Smith, M. K.), Chemistry Team (Stowe, R. L., Cooper, M. M.), Engineering Team (Prince, M.), Geography Team (Atit, K., Uttal, D. H.). "The Curious Construct of Active Learning." *Psychological Science in the Public Interest* 22 (1) (in press). https://doi.org/10.1177/1529100620973974

Lombardi, Doug, and Gale M. Sinatra. "College Students' Perceptions About the Plausibility of Human-Induced Climate Change." *Research in Science Education* 42, no. 2 (2012): 201–17.

Loria, Kevin. "These 18 Accidental and Unintended Scientific Discoveries Changed the World." Science Alert, April 4, 2018. https://www.sciencealert.com/these-eighteen-accidental-scientific-discoveries-changed-the-world.

Loveless, T. *Brown Center Report on American Education: How Well Are American Students Learning? Part III: A Progress Report on the Common Core.* Washington, DC: Brookings Institution, 2014.

Mann, Michael E. *The Hockey Stick and the Climate Wars: Dispatches from the Front Lines.* New York: Columbia University Press, 2013.

Matewos, Ananya M., Julie A. Marsh, Susan McKibben, Gale M. Sinatra, Q. Tien Le, and Morgan S. Polikoff. "Teacher Learning from Supplementary Curricular Materials: Shifting Instructional Roles." *Teaching and Teacher Education* 83 (2019): 212–24.

McIntyre, Lee. *The Scientific Attitude: Defending Science from Denial, Fraud, and Pseudoscience*. Cambridge, MA: MIT Press, 2019.

Mitchell, Fiona. "Vitamin-D and Covid-19: Do Deficient Risk a Poorer Outcome?" *Lancet Diabetes & Endocrinology* 8, no. 7 (2020): 570.

National Academies of Sciences, Engineering, and Medicine. *Science literacy: Concepts, contexts, and consequences*. Washington, DC: National Academies Press, 2016.

National Council on Teacher Quality. "Content Knowledge: Elementary Teacher Preparation Policy," May 2019a. https://www.nctq.org/yearbook/national/Content-Knowledge-75.

National Council on Teacher Quality. "Secondary Content Knowledge: Secondary Teacher Preparation Policy," April 2019b. https://www.nctq.org/yearbook/national/Secondary-Content-Knowledge-84.

National Research Council. *Learning Science in Informal Environments: People, Places, and Pursuits*. Washington, DC: National Academies Press, 2009.

National Science Foundation. "Science & Engineering Indicators 2018," 2018. https://www.nsf.gov/statistics/2018/nsb20181/.

National Science Teaching Association. "About the Next Generation Science Standards," 2014. https://ngss.nsta.org/About.aspx.

NGSS Lead States. *Next Generation Science Standards: For States, by States*. Washington, DC: National Academies Press, 2013.

Picheta, Rob. " 'Missing Link' in Human History Confirmed After Long Debate." CNN, January 19, 2019. https://www.cnn.com/2019/01/19/health/australopithecus-sediba-human-history-scli-intl/index.html.

Polikoff, Morgan, Q. Tien Le, Robert W. Danielson, Gale M. Sinatra, and Julie A. Marsh. "The Impact of Speedometry on Student Knowledge, Interest, and Emotions." *Journal of Research on Educational Effectiveness* 11, no. 2 (2018): 217–39.

Rudolph, John L. *How We Teach Science—What's Changed, and Why It Matters*. Cambridge, MA: Harvard University Press, 2019.

Schmidt, William H., Nathan A. Burroughs, Pablo Zoido, and Richard T. Houang. "The Role of Schooling in Perpetuating Educational Inequality: An International Perspective." *Educational Researcher* 44, no. 7 (2015): 371–86.

Sinatra, Gale M., and Suzanne H. Broughton. "Bridging Reading Comprehension and Conceptual Change in Science: The Promise of Refutation Text." *Reading Research Quarterly* 46, no. 4 (2011): 374–93.

Sinatra, Gale M., and Robert W. Danielson. "Adapting Evolution Education to a Warming Climate of Teaching and Learning." In *Evolutionary Perspectives on Child Development and Education*, edited by David C. Geary and Daniel B. Berch, 271–90. Cham, Switzerland: Springer, 2016.

Sinatra, Gale M., and Doug Lombardi. "Evaluating Sources of Scientific Evidence and Claims in the Post-Truth Era May Require Reappraising Plausibility Judgments." *Educational Psychologist* 55, no. 3 (2020): 120–31.

Sinatra, Gale M., Ananya Mukhopadhyay, Taylor N. Allbright, Julie A. Marsh, and Morgan S. Polikoff. "Speedometry: A Vehicle for Promoting Interest and Engagement through Integrated Stem Instruction." *Journal of Educational Research* 110, no. 3 (2017): 308–16.

Sinatra, Gale M., Sherry A. Southerland, Frances McConaughy, and James W Demastes. "Intentions and Beliefs in Students' Understanding and Acceptance of Biological Evolution." *Journal of Research in Science Teaching* 40, no. 5 (2003): 510–28.

Smithsonian Science Education Center. "The STEM Imperative," 2020. https://ssec. si.edu/stem-imperative.

Speedometry. "Math and Science Curriculum: STEM Lesson Plans & Activities," 2020. http://origin2.hotwheels.mattel.com/en-us/explore/speedometry/index.html.

Suldovsky, Brianne. "In Science Communication, Why Does the Idea of the Public Deficit Always Return? Exploring Key Influences." *Public Understanding of Science* 25 (2016): 415–26.

Thacker, Ian, Gale M. Sinatra, Krista R. Muis, Robert W. Danielson, Reinhard Pekrun, Philip H. Winne, and Marianne Chevrier. "Using Persuasive Refutation Texts to Prompt Attitudinal and Conceptual Change." *Journal of Educational Psychology* 112, no. 6 (2020): 1085–99.

Torney-Purta, Judith, and Jo-Ann Amadeo. "International Large-Scale Assessments: Challenges in Reporting and Potentials for Secondary Analysis." *Research in Comparative and International Education* 8, no. 3 (2013): 248–58.

Vosniadou, S., ed. *International Handbook of Conceptual Change*. New York: Routledge, 2008.

Vosniadou, S., and W. F. Brewer. "Mental Models of the Earth: A Study of Conceptual Change in Childhood." *Cognitive Psychology* 24 (1992): 535–85.

World Health Organization. "Food, Genetically Modified," 2020. https://www.who.int/ health-topics/food-genetically-modified/#tab=tab_1.

SECTION II

FIVE EXPLANATIONS
FOR SCIENCE DENIAL, DOUBT,
AND RESISTANCE

4

How Do Cognitive Biases Influence Reasoning?

"We need to educate our members about the health consequences of GMOs,"
Mark tells the other food co-op board members at their monthly meeting. "Our
members have pressed us to prohibit the sale of food from genetically modified
organisms, and they will be glad we have voted to do it. Now it's our job to make
sure all the members know why GMOs are unhealthy to eat." The board agrees
that Mark will write a brief piece for the co-op's newsletter. Like several others
at the meeting, his opposition to genetically modified organisms (GMOs) has
arisen from concerns about control of the seed supply by a few large corpor-
ations that are patenting their GMO seeds. He worries about how local farmers
will be affected, no longer allowed to save their own seeds and paying premium
prices for the modified ones. He is concerned about environmental risks and
the loss of biodiversity and that insects may be affected that are not the target
of the pesticides in the new breeds. GMOs can't be good for the human body, he
assumes, which aligns with what he has heard from those who oppose selling ge-
netically modified food at the co-op. They say they are concerned it will increase
obesity, alter human DNA, lead to autism, cause cancer, and create allergic re-
actions. Now he just needs to confirm the facts and educate the members, part
of the food co-op's mission.

Seeking information about the negative health effects of eating GMOs, he
starts by choosing search terms that lead him to supportive information (e.g.,
"are GMOs bad for health?"), confirming what he believes. He looks further, to
make sure he knows enough to write the article, and is surprised and uncom-
fortable to learn that the negative health consequences have been challenged
by scientists. He learns about "golden rice," a genetically modified strain that
produces beta-carotene and could prevent blindness in the developing world.
Moreover, some of the health concerns members have about GMOs appear to be
attributed to pesticides, not GMOs. He discovers that the issue is far more com-
plex than he had imagined and that various governments, including that of the
European Union, have altered their stance over time; the European Union first

Science Denial. Gale M. Sinatra and Barbara K. Hofer, Oxford University Press. © Oxford University Press 2021.
DOI: 10.1093/oso/9780190944681.003.0004

banned GMOs, then allowed countries to decide. He learns that the US govern-ment supports labeling but then wonders how consumers are supposed to use that information.

How will Mark evaluate the science he reads? Will he selectively pick the readings that support his beliefs or examine them less critically? What are the mental processes that might impede his ability to get an objective, unbiased perspective on a complicated socio-scientific issue such as this one? How do they operate in the background of his awareness? How do each of us learn to mitigate the fallacies of our own reasoning systems?

Understanding Our Own Thinking

We all like to think of ourselves as rational actors, careful and considered in our thinking, capable of sound and reliable judgments. We might believe that we generally consider different points of view and make informed decisions. We are, in fact, "predictably irrational," as psychologist Dan Ariely titled his book on the topic.[1] All of us engage in automatic, reflexive thinking, typically taking the easier path and conserving mental effort. Although we each may have the subjective impression that we are careful thinkers, we often make snap judgments or no real judgments at all. In addition, numerous biases inhibit or override reflective, deliberative thought; intuitive theories can also impede ac-ceptance of accurate scientific explanations. Understanding more about how our minds work and how biases may operate can make us each less suscep-tible to fallacious reasoning, more rational, and more aware of the problems in others' thinking. Learning to understand the built-in limitations of our mental processes can also help us improve our ability to inform others more effectively.

Thinking Fast and Slow: System 1 and System 2

Psychologists have described a dual-processing aspect to our brains, the experiential and the analytical.[2] The more primitive part of the brain that reacts quickly, making swift intuitive decisions, drawing on experience and emotion, is what Nobel Prize–winning social psychologist Daniel Kahneman calls "System 1."[3] We each frequently start from a decision or a belief, and then stories arise in our mind that explain what we think—so conclusions come first, and rationalizations follow.[4] (e.g., System 1 thinking might be "I don't want to see my business shuttered because of the pandemic, and I heard that COVID-19 is just like the flu—and my neighbor didn't get all that sick when she had it.") It takes cognitive effort to turn on the more analytical, reflective, deliberative mind of System 2, the logical, slower, and more thoughtful part of thinking. ("But when I read what the medical experts are saying and look at the data, I see that this is more serious and deadly than the flu, and I need to take precautions—and accept that we might not do business as normal for a while.") Although some have criticized the duality as oversimplified,[5] the dual-processing model illustrates a reflexive approach to decision-making and the work needed to override it.

The brain has been described by social psychologists Susan Fiske and Shelly Taylor as a "cognitive miser," responding in energy-efficient ways to the vast amount of information received.[6] As a result, humans often take mental shortcuts, thinking as quickly and simply as possible. These processes have evolutionary origins, and lightning-quick decisions were useful in escaping predators in the ancestral environment of the savanna, enhancing survival. Fast, frugal, automatic responses continue to help preserve cognitive energy, and no one would want to spend time carefully deliberating all the decisions required in a single day. But the System 1 mechanism is woefully inadequate for the complexities of modern living, especially in understanding science or making socio-scientific decisions. Quick judgments doom individuals to choices that may feel intuitively and emotionally right but are often quite wrong, and they can lead to stereotypes and prejudicial thinking.

Defaulting to System 1, without the checks and balances of System 2, can make one susceptible to manipulation by others. For example, individuals are more responsive to information presented in an emotional way, which has become the basis for some political messaging, fear-mongering from anti-vaccination groups, and the persuasive attempts of others who wish to influence decisions and behaviors. Bombarded by emotion-inducing material, individuals can learn to recognize the signs of a cognitive minefield and call deliberately on the rationality of System 2.

System 2 thinking is effortful and more time-consuming, however; it would be exhausting and inefficient to employ it consistently. Learning when to slow thinking down, and learning how to do it, is essential for informed decision-making and scientific thinking. As Kahneman notes,

"Little can be achieved without a considerable investment of effort."[7] System 2 allows one to examine cognitive errors and bring them into check, marshal relevant evidence, engage in analysis, and move toward effective decisions and action.

Engaging System 2 does not produce flawless thinking. People do not always operate with good data or have access to the full resources needed for wise decision-making. Sometimes System 2 endorses or rationalizes the judgments of System 1. As cognitive misers, individuals might also be willing to accept limited evidence or biased data. The development of effective science understanding involves monitoring System 1 impulses plus a willingness to expend the energy to employ System 2 thoroughly, with more than a brief review of those initial impulses. Note, however, that System 1 also has benefits; any skilled performance relies on the automatic processing of information that happens fast and efficiently.

Confidence by Coherence

Another facet of dual-processing minds is that "The sense-making machinery of System 1 makes us see the world as more tidy, simple, predictable, and coherent than it really is."[8] When Mark volunteered to educate co-op members about the negative health consequences of eating genetically modified organisms, he made one of those quick mental leaps, assuming coherence among a set of beliefs. Surely the political, agricultural, and corporate critiques of GMOs that he knew well should align with problematic health outcomes, which would be consistent and congruent with the overall message the co-op board believed, that GMOs are bad.

Similarly, without understanding the effects and mechanisms of climate change, one might think "global warming" means that all places on the planet are *consistently* getting warmer, rather than recognizing that swings and extremes in local temperatures are more likely, with an overall progression toward a longer-term warming trend in the climate. The idea seems logically coherent, however unfounded. Moreover, because this view is also contradicted by experience, individuals then become dismissive of the idea of a warming trend. In February 2015, climate denialist US Senator Jim Inhofe, author of *The Greatest Hoax: How the Global Warming Conspiracy Threatens Your Future*,[9] tossed a snowball in the Senate chambers, supposedly demonstrating the myth of climate change to fellow members of Congress. How could global warming exist if Washington, DC, had record cold and snow? Even in 2019, with the catastrophic effects more evident and a national scientific report calling for immediate action, the US president tweeted, "In the beautiful Midwest, windchill temperatures are reaching minus 60 degrees, the coldest ever recorded. . . . What the hell is going on with Global Waming [*sic*]? Please come back fast, we need you!"

Such subjective confidence in one's own experience can reflect what Kahneman calls "confidence by coherence." Even poor evidence—a snowball in February or record-setting cold temperatures—can make a very good story. Beliefs held with great confidence, however, are not necessarily true. Our System 1 minds prefer the suppression of doubt and evoke ideas that are compatible with the stories we are telling ourselves. As the pandemic raged across the United States, some individuals were literally denying to the grave, so convinced that the virus was a hoax that even on their deathbeds they were proclaiming it must be something else infecting them.

Sometimes people have no evidence at all for beliefs they hold dear, but people they trust hold those particular beliefs, and that seems adequate. Believing the ideas of those you love and trust is a form of coherence. In the research that we, the authors, have done with college students regarding their beliefs about evolution, some students prefer creationism over evolution because it is consistent with the religious beliefs they have been taught and consistent with what their families believe. College science classes that provide scientific accounts of evolution can be challenging for those expecting coherence. Students need support in untangling the complexity of these educational issues and in learning to acknowledge and value scientific evidence.

Intuitive Theories: Are People Scienceblind?

The need for coherence can reinforce one's own intuitive beliefs about scientific phenomena, and many harbor a good number of theories that help them explain science to themselves, no matter how erroneous they might be. Try thinking about the following scenario and consider your own confidence in the answer. Imagine yourself with two bullets: you place one in the chamber of a gun, and you hold the second in your other hand. If you were standing in a large open field and dropped the second bullet just as you pulled the trigger on the gun, which bullet would hit the ground first?

As cognitive psychologist Andrew Shtulman explains in *Scienceblind: Why Our Intuitive Theories About the World Are So Often Wrong*, most people will answer incorrectly, believing that a forward-propelling force would keep the shot bullet aloft longer. Gravity, however, is the single force that brings both bullets to the ground at the same time, although one will have traveled further. The explanation for such faulty reasoning, backed by decades of psychological research, is that individuals default to their intuitive theories about how the world works. In this case, an intuitive theory of motion involves the belief that objects move because of an imparted force or impetus, dissipating over time,[10] and that projectiles have forces imparted to them. Following that erroneous logic, something that is dropped has no imparted force to keep it in motion the way the bullet propelled from the gun does.

On a cognitive level, intuitive theories are different from erroneous knowledge and misinformation, which can more readily be corrected. A child who has only ridden in cars fueled by gas may be surprised to learn that the new family car needs to be plugged in when parked in the garage, but such knowledge of how cars run can be easily adjusted when the parents explain. By contrast, misconceptions based on the idea of imparted force are common worldwide and across the age span, and such intuitive theories are more coherent and resistant to change. Moreover, such deeply held naive theories exist in a wide variety of scientific areas, as Shtulman illustrates: matter, life, energy, illness, and scientific topics such as climate change, evolution, and continental drift. These untutored explanations are our best guesses of how the world works. The problem is that they can also become deeply held beliefs that may close our minds to competing explanations.

Most US citizens know that in spite of the perception that the sun "rises" each day, Earth is in fact spinning around the sun. (Surprisingly, 27% of US citizens queried in a National Science Foundation study answered that question incorrectly.[11]) Yet when asked to consider a related but slightly more complex question, why Earth has seasons, most adults will resort to faulty explanations about an elliptical orbit, rather than describing the effects of Earth's tilt on its axis. (Earth's orbit around the sun is nearly circular.) Research has shown that it takes serious mental effort for individuals to change their naive ideas or to suppress formerly held naive conceptions and give scientific responses. Even when individuals learn a correct scientific answer, they are slower to provide it when it contradicts an earlier intuitive theory, suggesting that they never fully displace that early belief and have to work to override it.[12] In some cases, experience may contradict the scientific findings, as in the case of seeing the sun appear to rise, even if you know that is not actually true; and it is also easy to imagine why the idea of a flat earth endured and was difficult to rattle.

Science learning is especially difficult when a scientific perspective conflicts with one's own naive ideas, and robust intuitive theories can be especially change-resistant. Imagine yourself sitting in a completely dark room with no light at all. Can you still see anything?

In the clip "Can We Believe Our Eyes?" from the film *A Mind of One's Own*, researchers at the Harvard Center for Astrophysics interview middle school students on this topic, who seem certain, incorrectly, that they could see in absolute darkness. Seated at a table with an apple in front of her, Karen elaborates, "But it wouldn't be red. It would just be gray or a lighter shade of black." When they try it, sealing the room to create darkness, she acknowledges she cannot see the apple, even after five or six minutes, "So I guess I was wrong," she laughs. Yet, when pressed, she does not fully accept what she has experienced, clinging to her erroneous ideas. "I think eventually your eyes will adjust. It might take a couple years."

Intuitive theories also exist for psychological phenomena and can be equally difficult to rattle. Imagine the college student who struggles in a new course. He decides that the problem is the instructor's methods, which just don't fit his "learning style," an explanation he has often generated in these situations. Many college students hold such beliefs—first, that they have a definitive learning style and, second, that they learn best when instructors match their teaching to that learning style. This is not surprising since this is a widely held belief among K–12 teachers, who find it useful in understanding learner differences. But in spite of a significant number of studies that show it is simply not the case that such a "match" is educationally beneficial,[13] students continue to harbor such beliefs, often to the detriment of their own learning. They may externalize blame when they don't do well, instead of figuring out the appropriate learning strategies for a particular course or subject or by seeking help in office hours. In our undergraduate educational psychology courses, we (the authors) have found that it takes multiple readings and discussion of research that show why their naive beliefs about learning styles are not supported by evidence, what psychologists call "refutational readings."[14] But it also requires helping them understand why scientific evidence is a more useful tool than their own intuition for understanding how learning occurs and that it can be an impediment to their learning to harbor erroneous ideas that are unfounded.[15]

Regardless of the discipline or the concept, Shtulman advocates tutorials that challenge learners' theories and help them refine or elaborate them. Students need to understand their own theories, the explanation of why they are flawed, and how scientific theories of the same phenomena fare better.[16] However, often individuals are not motivated to take the time and effort to be reflective and to carefully evaluate the science that supports non-preferred conclusions. Such efforts may be necessary to evaluate scientific conclusions fairly. In one study, individuals were asked to quickly verify the accuracy of statements that were either consistent or inconsistent with scientific theories.[17] Under time pressure, individuals were quicker to confirm statements that were accurate than the ones that were scientifically inaccurate. What the researchers argued is that individuals, when pressed to answer quickly, have to take an extra step to inhibit the tendency to revert back to their original naive conceptions. Yet how frequently are we each likely to make snap judgments about what we accept as true? Taking time to reflect on veracity is critical.

Thinking Anecdotally Versus Thinking Scientifically

Humans are wired to think anecdotally, but they can learn to think scientifically. As we saw with System 1 and System 2, we all favor experience, both our own and that of others, through the stories they tell us, over analysis. Perhaps you have a neighbor who is passionate about the new car he bought, and you decide to buy the same car, ignoring the reviews you read that could lead to more informed, wiser choices, grounded in data. Or maybe your child has a teacher who is concerned that he doesn't always do his homework and, remembering what her teachers did, keeps him in at recess, in spite of all she learned in her college classes about why such punishment is not likely to be effective in changing this behavior. Perhaps you have a relative who got sick a week after she received a flu shot, and she is no longer persuaded that vaccinations work, contrary to scientific evidence, and decides not to bother with them anymore—and tries to persuade you to do the same. Examples abound, suggesting just how difficult it is to override the experiential, anecdotal aspects of our thinking.

It takes more effort to learn to favor scientific theories over intuitive theories—and our world depends on it. Learning to prioritize science knowledge over intuition is a critical component of science education. This type of education requires more than exposure to knowledge that can be repeated on tests; it also requires an understanding of how science is conducted and learning to think like a scientist.

Illusion of Understanding

How confident are you that you understand what causes climate change, on a 1 (not at all confident) to 5 (very confident) scale? Quickly, take a minute, and explain the mechanism to yourself. Was this harder to do than you thought?

In a series of five studies, psychologists Michael Ranney and Dav Clark at the University of California, Berkeley, found very few individuals they surveyed, even those with considerable scientific literacy, who could describe the mechanism of anthropogenic (human-caused) global warming with even basic accuracy.[18] They explain the mechanism this way: "Earth transforms sunlight's visible light energy into infrared light energy, which leaves Earth slowly because it is absorbed by greenhouse gases. When people produce greenhouse gases, energy leaves Earth even more slowly—raising Earth's temperature."[19] How correct was your own explanation?

Perhaps you were not confident that you knew the causes of climate change, but in general individuals often misjudge the depth of their own understanding, imagining they know more than they do. This *illusion of understanding* is prevalent, especially with complex topics such as GMOs, climate change, or stem cell research. Confidence in one's explanatory knowledge often exceeds what is actually understood, whether that is combination locks, refrigerators, toilets,[20] or more complex and critically important topics. Not only are most people incapable of generating accurate explanations, but they are unaware of how little they actually grasp. What Ranney and Clark have shown is that explanatory, mechanistic knowledge can matter in our acceptance of scientific ideas and that it is teachable. In two other experiments, they found that teaching the mechanism of climate change reduced the illusion of explanatory depth and increased climate change acceptance—notably, across the political spectrum.[21] Buoyed by this success, they launched a website designed to help the public learn the mechanistic explanations for climate change, HowGlobalWarmingWorks.org.[22] The website allows the viewer to

choose from a range of videos to watch, from a less than one-minute explanation to a five-minute, more detailed version. Their research with these simple
video interventions shows changes in acceptance.[23]

Understanding the basic mechanisms of topics such as climate change can
be critical to accepting the scientific consensus on such a politically charged
topic, but no one can understand the scientific explanations of everything in
the environment. Given the complexity of human inventions over time and
the complicated science that underlies many natural phenomena, this just
is not possible, as psychologist Steven Sloman and cognitive scientist Philip
Fernbach argue in their book *The Knowledge Illusion: Why We Never Think
Alone*.[24] What one needs to discern is when it matters to health and well-
being, as well as to others (e.g., vaccinations, stem cell research) and to the
planet (e.g., climate change), and when to make a concerted effort to challenge these illusions of understanding. Ignorance of mechanisms can impede
acceptance of scientific conclusions on critical scientific topics—or create
ungrounded fears. Illusions of explanatory depth can also explain extremism
in political attitudes, yet this too can be addressed, with notable effects. When
individuals were asked to state their positions on a set of six policies (e.g.,
imposing sanctions on Iran, establishing a cap-and-trade policy for carbon
emissions, transitioning to single-payer healthcare) and then asked to provide a mechanistic explanation for one of the policies, their attitudes became
more moderate.[25]

A frequently cited quote from Donald Rumsfeld, US Secretary of Defense
for both Gerald Ford and George W. Bush, has often been parodied: "There
are known knowns. There are things we know that we know. There are
known unknowns. But there are also unknown unknowns. There are things
we don't know that we don't know." As humorous as this musing about knowledge might be, we would all do better in addressing the illusion of understanding by beginning to recognize what we don't actually know. The fact
that anyone can so quickly access any information needed from smartphones
and computers may further complicate the problem. They may be exacer-
bating the illusion by thinking they don't actually need to know because
they can always just look it up. This complacent ignorance can lead to what
Nicholas Carr warns about in *The Shallows: What the Internet Is Doing to Our
Brain*.[26] He envisions a population with knowledge too shallow to perform
the higher-order thinking functions of analyzing, synthesizing, and creating.
Unfortunately, this will not help solve the socio-scientific problems that face
individuals and the larger community.

Confirmation Bias

In the opening vignette, Mark's co-op board made the decision to ban genetically modified organisms—and then Mark looked for evidence to support the board's existing position that such foods are unhealthy to eat. We each engage in such *confirmation bias*[27] when we seek, interpret, or recall information that aligns with preexisting beliefs. Basically, we are more likely to pay attention to information we already believe is true. It requires vigilance to guard against this mental tendency, to stay open to new perspectives, and to evaluate information that challenges what you think you know—or want to believe is true. It is much simpler to accept the information that supports what you already think and ignore what contradicts your beliefs. Moreover, with the current filter bubbles present in the internet era (where algorithms feed us information that fits our existing worldviews), individuals might not even see information that challenges preconceived notions, unless they are vigilant in seeking it out.

Humans are also less likely to see errors in their own thinking than errors in others' thinking, displaying another form of what cognitive scientist Hugo Mercier prefers to call "myside bias."[28] In one of his experiments, conducted with colleagues, they found that students produced more stringent arguments when critiquing the work of others than when critiquing arguments they had produced on their own. In an intriguing twist, one of the studies involved reviewing arguments participants were told others had generated but were actually their own, produced in an earlier part of the study. Those who didn't recognize the source of the arguments as self-generated were more likely to reject the arguments. This appears to be one more example of what the authors deem "selective laziness." Favoring an evolutionary model, Mercier and Dan Sperber note that the bias toward one's own position might have been helpful in another era but that the environment has been changing too quickly for this to be a beneficial response in the current time.[29]

Confirmation bias is further influenced by emotion. Beliefs that are generated during times of strong emotion are likely to be more resistant to disconfirmation, what behavioral psychologist Paul Slovic has called the "affect heuristic."[30] For example, someone with strong fears about defending themselves against a potential burglar and who keeps a gun by the bed may arouse that fear when gun control is discussed. They may erroneously believe that it is safer to keep a gun at home, in spite of evidence to the contrary. Their emotional response to the topic may make it difficult to accept the substantial body of research that contradicts that position. A 2014 review of existing studies on the topic in the *Annals of Internal Medicine* shows that access to firearms (even when properly stored) increases the risk of violence in the home.[31]

Availability Heuristic

Other reasoning flaws can further impede clear thinking. For example, we are all easily prone to the *availability heuristic*, the tendency to access whatever information is most available and relevant to us. For example, well-publicized events like plane crashes tend to magnify a sense of their likelihood. This mental shortcut predisposes individuals to the information they can quickly recall as a foundation for beliefs. We each typically give greater credence to what we can readily access and then overestimate its probability; thus, the weather one has experienced may affect one's ideas about climate change occurrence. The availability heuristic also makes us open to influence by those who want particular ideas to become salient in our minds and can make us susceptible to erroneous information. Senator Inhofe's display of the snowball in Washington as evidence against global warming shows how memorable such an example can be. When political leaders chose not to wear masks during the outbreak of the global pandemic, even as health officials were encouraging it, many citizens retained this as evidence it was not necessary to do so themselves.

Thinking Dispositions

As you read the examples in the previous sections of this chapter, did you take time to work them out yourself, or did you skip over that part? Were you curious about your own thinking and beliefs? Were you willing to expend effort to understand them better? In addition to operating with cognitive biases and heuristics, we each have what psychologist Keith Stanovich calls "thinking dispositions." These dispositions influence our willingness to engage in rational thinking, exert cognitive energy, challenge our own beliefs, and operate with an open mind. When educators talk about critical thinking, they are often referring to these malleable processes. These processes involve learning to evaluate arguments and evidence in ways not contaminated by prior beliefs.[32]

Imagine that our protagonist Mark, after researching whether GMO foods are unhealthy to eat, returns to the co-op board with a slew of well-organized information that contradicts what the board members initially believed. How will the board members respond? Mark may need to be prepared to witness a range of individual differences that may reflect these thinking dispositions. Will they spring into rebuttal, or will they weigh new evidence against their preferred belief? Will the members invest the effort to hear alternative viewpoints with an open mind? Stanovich and colleague Richard West[33] investigate what they call "actively open-minded thinking" by examining the degree to which individuals agree with statements such as "People should always take into consideration evidence that goes against their own beliefs" and disagree with items such as "No one can talk me out of something I know is right." Individuals who hold truth as a value and who are willing to change their beliefs to get closer to the truth are going to be more open to hearing Mark's findings. These individuals are more likely to seek accuracy in their beliefs and have the cognitive flexibility to remain open to new information. Individuals can examine whether they are more likely to reason in a data-driven manner or a belief-driven way.[34]

Another individual difference in thinking is *need for cognition*, a preference for effortful cognitive activities.[35] Need for cognition is measured with items such as "I would prefer a task that is intellectual, difficult, and important to one that is somewhat important but does not require much thought." As college professors, we see this manifest in our work with students, at both the undergraduate and graduate levels. Some are happy to have simple, surface-level information on course topics and move forward blithely, while others dig deeper, read outside the required list, come to office hours and ask hard questions, and work to make complex information intelligible, beyond what is expected for their assessed performance. While we may delight in students who arrive with this propensity for difficult cognitive endeavors, our job is to also fuel such intellectual curiosity in others and to model it where we can, striving to foster habits of mind that will promote deeper learning. The connection to science understanding is also clear. High need for cognition correlates with both deliberate thinking and effortful, high-quality evaluation of evidence. Low need for cognition is less beneficial; in one study, individuals low in need for cognition were less likely to notice scientific explanations that were circular (assuming what is intended to be proven).[36]

What Can We Do?

How might Mark find his way through the thicket of competing explanations about the safety of GMOs and learn to understand the science? How does he overcome his own biases? How does he encourage his fellow board members to truly listen in ways that help them entertain complexity and nuance? How does he write his article for co-op members in a way that allows readers to be open to understanding beyond their initial beliefs? How does he use good science communication tactics to educate the members? More importantly, what can each of us do, knowing how easy it is to default to what we already believe? What can educators do? There are no easy solutions, but here are some tips.

What Can Individuals Do?

Learning to do the effortful work of engaging System 2 thinking and becoming aware of one's own biases and heuristics become paramount when interpreting complex scientific topics.

> *Practice slowing down and allowing for a more thoughtful, informed response.* As Kahneman notes, the automatic processor of System 1 does not send a warning signal when it becomes unreliable, nor does Kahneman think it is readily educable.[37] The best way to block System 1 errors is to recognize when you are in a cognitive minefield and seek the help of System 2. When you find yourself slipping into quick, pre-programmed thinking about a scientific (or other) issue of meaning and complexity, back off for a moment. Consider other possibilities. Engage your analytical mind, take the time to seek out and evaluate evidence, and allow for other conclusions. Moreover, when a topic is important to you, make sure you take the time to check with yourself about whether you are skipping these steps.
>
> *Watch for your own confirmation bias, often easy to spot during online searches, once awareness is turned on.* If you already think you know the

answer (to a health issue, to a policy concern, to a controversial science issue), do you stop searching when you find that response? Check that impulse, dive deeper, consider the sources of information, and look for empirical evidence from reputable sources.

In discussions with others, you can practice keeping an open mind, remembering that learning why others think as they do doesn't mean accepting what they believe. You can communicate scientific information with clear narratives and evidential support. For example, Mark can hear his colleagues' concerns about GMOs, provide refutation both orally and in print, and then offer to have another meeting after the board has had time to read the material and evaluate its sources. He can listen to what their concerns are and help them disentangle their beliefs, helping them avoid confidence by coherence. As a set of four studies has shown, learning the science behind basic genetic modification technology leads to more positive attitudes toward foods that have been genetically modified and lower perception that they are risky to eat.[38] The work of Gale and her colleagues shows that when individuals develop more scientifically accepted views of genetically modified organisms, the more positive emotional shift that follows predicts more positive attitudes.[39] Mark's coop board members may persist in their reasoned objections to the impact that GMOs have on farming, but might also pursue more information about whether there are health risks to consumption.

What Can Educators Do?

If you are an educator, you can encourage students to think scientifically, to question their own intuitive responses, and to strive toward greater scientific understanding.

Foster thinking dispositions that press students to think more deeply and analytically, by providing opportunities for practice and by nurturing curiosity and open-mindedness. Help them overcome illusions of understanding by asking them to explain the phenomenon under study and then work with them to develop a coherent, reliable explanation consistent with scientific information. Shtulman notes that students need to learn not just scientific facts but also to think like scientists in order to replace intuitive theories with scientific theories. He advocates introducing science as a method of reasoning, not just a set of solutions.[40]

Consider creating science assignments that invite students to challenge their own assumptions, intuitions, and prior knowledge. Help them build a bridge from what they think they know to deeper, richer understandings that might contradict initial impulses. Give students practice in the evaluation of evidence. Assist them in developing awareness of the desire to confirm what they already believe. While it would be unreasonable (or even undesirable) to assume that they will always slow down their thinking and employ more cautious, analytical reasoning, you can help students develop the conditional, self-regulatory knowledge to know when it matters.

Give students practice in spotting logical errors in others' thinking and argumentation, as a means of learning to recognize it in their own reasoning. Help them gain awareness about the ways in which emotion might impair their judgments. Provide rich examples that help them see the flaws in thinking that this chapter illustrates and encourage them to identify their own such faulty patterns and the means for correcting them. None of these endeavors will be successful unless students are also motivated and engaged, critical components of education.

Conclusions

Becoming aware of biases, heuristics, blind spots, and the quick responses of our System 1 thinking is the first step in becoming more vigilant in one's assessment of information, as well as in countering erroneous conclusions from others. Such training needs to be a part of educational efforts alongside strong science education.

Notes

1. Dan Ariely, *Predictably Irrational: The Hidden Forces That Shape Our Decisions* (New York: Harper Perennial, 2008).
2. Eric Amsel et al., "A Dual-Process Account of the Development of Scientific Reasoning: The Nature and Development of Metacognitive Intercession Skills," *Cognitive Development* 23 (2008).
3. Daniel Kahneman, *Thinking, Fast and Slow* (New York: Farrar, Straus and Giroux, 2011).

4. Krista Tippet, "Interview with Daniel Kahneman," *On Being*, October 7, 2017.

5. David E. Melnikoff and John A. Bargh, "The Mythical Number Two," *Trends in Cognitive Science* 22 (2018).

6. Susan T. Fiske and Shelley E. Taylor, *Social Cognition* (New York: McGraw-Hill, 1984).

7. Kahneman, *Thinking, Fast and Slow*, p. 204.

8. Kahneman, *Thinking, Fast and Slow*, p. 204.

9. James Inhofe, *The Greatest Hoax: How the Global Warming Conspiracy Threatens Your Future* (Hawthorne, NV: WND Books, 2012).

10. Peter A. White, "The Impetus Theory of Judgments About Object Motion: A New Perspective," *Psychonomic Bulletin and Review* 19 (2012).

11. National Science Foundation, *Science and Engineering Indicators 2018* (Alexandria, VA: National Science Foundation, 2018).

12. Andrew Shtulman and Kelsey Harrington, "Tensions Between Science and Intuition Across the Lifespan," *Topics in Cognitive Science* 8 (2016).

13. Beth A. Rogowsky, Barbara M. Calhoun, and Paula Tallal, "Matching Learning Style to Instructional Method: Effects on Comprehension," *Journal of Educational Psychology* 107 (2014). Joshua Cuevas, "Is Learning Styles–Based Instruction Effective? A Comprehensive Analysis of Recent Research on Learning Styles," *Theory and Research in Education* 13 (2015).

14. Gale M. Sinatra and Suzanne H. Broughton, "Bridging Reading Comprehension and Conceptual Change in Science: The Promise of Refutation Text," *Reading Research Quarterly* 46, no. 4 (2011).

15. Barbara K. Hofer, "Epistemic Cognition: Why It Matters for an Educated Citizenry and What Instructors Can Do," *New Directions for Teaching and Learning* (2019). "Shaping the Epistemology of Teacher Practice Through Reflection and Reflexivity," *Educational Psychologist* 52 (2017).

16. Andrew Shtulman, *Scienceblind: Why Our Intuitive Theories About the World Are So Often Wrong* (New York: Basic Books, 2017).

17. Andrew Shtulman and Joshua Valcarcel, "Scientific Knowledge Suppresses but Does Not Supplant Earlier Intuitions," *Cognition* 124 (2012).

18. Michael A. Ranney and Dav Clark, "Climate Change Conceptual Change: Scientific Information Can Transform Attitudes," *Topics in Cognitive Science* 8 (2016).

19. Michael A. Ranney, Daniel Reinholz, and Lloyd Goldwasser, "How Does Climate Change ('Global Warming') Work? The Mechanism of Global Warming, an Extra Greenhouse Effect," http://www.HowGlobalWarmingWorks.org.

20. Leonid Rosenblit and Frank Keil, "The Misunderstood Limits of Folk Science: An Illusion of Explanatory Depth," *Cognitive Science* 26 (2002).

21. Ranney and Clark, "Climate Change Conceptual Change."

22. Ranney, Reinholz, and Goldwasser, "How Does Climate Change ('Global Warming') Work?"

23. Ranney and Clark, "Climate Change Conceptual Change."

24. Steven A. Sloman and Phillip Fernbach, *The Knowledge Illusion: Why We Never Think Alone* (New York: Riverhead, 2017).

25. Ranney and Clark, "Climate Change Conceptual Change."

26. Nicholas G. Carr, *The Shallows: What the Internet Is Doing to Our Brains* (New York: Norton, 2010).

27. Raymond S. Nickerson, "Confirmation Bias: A Ubiquitous Phenomenon in Many Guises," *Review of General Psychology* 2 (1998).

28. Hugo Mercier, "Confirmation Bias—Myside Bias," in *Cognitive Illusions: Intriguing Phenomena in Judgement, Thinking and Memory*, ed. R. F. Pohl (New York: Routledge, 2014).

29. Hugo Mercier and Dan Sperber, *The Enigma of Reason* (Cambridge, MA: Harvard University Press, 2017).

30. Paul Slovic et al., "The Affect Heuristic," *European Journal of Operational Research* 177, no. 3 (2007).

31. Andrew Anglemyer, Tara Horvath, and George Rutherford, "The Accessibility of Firearms and Risk for Suicide and Homicide Victimization Among Household Members: A Systematic Review and Meta-Analysis," *Annals of Internal Medicine* 160 (2014).

32. Keith E. Stanovich, "Rationality, Intelligence, and Levels of Analysis in Cognitive Science," in *Why Smart People Can Be So Stupid*, ed. R. J. Sternberg (New Haven: Yale University Press, 2002).

33. Keith E. Stanovich and Richard F. West, "Reasoning Independently of Prior Belief and Individual Differences in Actively Open-Minded Thinking," *Journal of Educational Psychology* 89 (1997).

34. Priti Shah et al., "What Makes Everyday Scientific Reasoning So Challenging?" in *Psychology of Learning and Motivation* (Academic Press, 2017).

35. John T. Cacioppo and Richard E. Petty, "The Need for Cognition," *Journal of Personality and Social Psychology* 42 (1982).

36. Jillian Minahan and Karen L. Siedlecki, "Individual Differences in Need for Cognition Influence the Evaluation of Circular Scientific Explanations," *Personality and Individual Differences* 99 (2016).

37. Kahneman, *Thinking, Fast and Slow*.

38. Jonathon McPhetres et al., "Modifying Attitudes About Modified Foods: Increased Knowledge Leads to More Positive Attitudes," *Journal of Environmental Psychology* (2019).

39. Benjamin C. Heddy et al., "Modifying Knowledge, Emotions, and Attitudes About Genetically Modified Foods," *Journal of Experimental Education* 85 (2017).

40. Shtulman, *Scienceblind*.

References

Amsel, Eric, Paul A. Klaczynski, Adam Johnston, Shane Bench, Jason Close, Eric Sadler, and Rick Walker. "A Dual-Process Account of the Development of Scientific Reasoning: The Nature and Development of Metacognitive Intercession Skills." *Cognitive Development* 23 (2008): 452–71.

Anglemyer, Andrew, Tara Horvath, and George Rutherford. "The Accessibility of Firearms and Risk for Suicide and Homicide Victimization Among Household Members: A Systematic Review and Meta-Analysis." *Annals of Internal Medicine* 160 (2014): 101–10.

Ariely, Dan. *Predictably Irrational: The Hidden Forces That Shape Our Decisions.* New York: Harper Perennial, 2008.

Cacioppo, John T., and Richard E. Petty. "The Need for Cognition." *Journal of Personality and Social Psychology* 42 (1982): 116–31.

Carr, Nicholas G. *The Shallows: What the Internet Is Doing to Our Brains.* New York: Norton, 2010.

Cuevas, Joshua. "Is Learning Styles–Based Instruction Effective? A Comprehensive Analysis of Recent Research on Learning Styles." *Theory and Research in Education* 13 (2015): 308–33.

Fiske, Susan T., and Shelley E. Taylor. *Social Cognition.* New York: McGraw-Hill, 1984.

Heddy, Benjamin C., Robert W. Danielson, Gale M. Sinatra, and Jesse Graham. "Modifying Knowledge, Emotions, and Attitudes About Genetically Modified Foods." *Journal of Experimental Education* 85 (2017): 513–33.

Hofer, Barbara K. "Shaping the Epistemology of Teacher Practice Through Reflection and Reflexivity." *Educational Psychologist* 52 (2017): 299–306.

Hofer, Barbara K. "Epistemic Cognition: Why It Matters for an Educated Citizenry and What Instructors Can Do." *New Directions for Teaching and Learning* 164 (2020): 85–94.

Inhofe, James. *The Greatest Hoax: How the Global Warming Conspiracy Threatens Your Future.* Hawthorne, NV: WND Books, 2012.

Kahneman, Daniel. *Thinking, Fast and Slow.* New York: Farrar, Straus and Giroux, 2011.

McPhetres, Jonathon, Bastiaan Rutjens, Netta Weinstein, and Jennifer Brisson. "Modifying Attitudes About Modified Foods: Increased Knowledge Leads to More Positive Attitudes." *Journal of Environmental Psychology* 64 (2019): 21–29.

Melnikoff, David E., and John A. Bargh. "The Mythical Number Two." *Trends in Cognitive Science* 22 (2018): 280–93.

Mercier, Hugo. "Confirmation Bias—Myside Bias." In *Cognitive Illusions: Intriguing Phenomena in Judgement, Thinking and Memory*, edited by R. F. Pohl, 99–114. New York and London: Routledge, 2014.

Mercier, Hugo, and Dan Sperber. *The Enigma of Reason.* Cambridge, MA: Harvard University Press, 2017.

Minahan, Jillian, and Karen L. Siedlecki. "Individual Differences in Need for Cognition Influence the Evaluation of Circular Scientific Explanations." *Personality and Individual Differences* 99 (2016): 113–17.

National Science Foundation. *Science and Engineering Indicators 2018.* Alexandria, VA: National Science Foundation, 2018.

Nickerson, Raymond S. "Confirmation Bias: A Ubiquitous Phenomenon in Many Guises." *Review of General Psychology* 2 (1998): 175–220.

Ranney, Michael A., and Dav Clark. "Climate Change Conceptual Change: Scientific Information Can Transform Attitudes." *Topics in Cognitive Science* 8 (2016): 49–75.

Ranney, Michael A., Daniel Reinholz, and Lloyd Goldwasser. "How Does Climate Change ('Global Warming') Work? The Mechanism of Global Warming, an Extra Greenhouse Effect." http://www.HowGlobalWarmingWorks.org.

Rogowsky, Beth A., Barbara M. Calhoun, and Paula Tallal. "Matching Learning Style to Instructional Method: Effects on Comprehension." *Journal of Educational Psychology* 107 (2014): 64–78.

Rosenblit, Leonid, and Frank Keil. "The Misunderstood Limits of Folk Science: An Illusion of Explanatory Depth." *Cognitive Science* 26 (2002): 521–62.

Shah, Priti, Audrey Michal, Amira Ibrahim, Rebecca Rhodes, and Fernando Rodriguez. "What Makes Everyday Scientific Reasoning So Challenging?" In *Psychology of Learning and Motivation*, edited by B. Ross, 251–99. Cambridge, MA: Academic Press, 2017.

Shtulman, Andrew. *Scienceblind: Why Our Intuitive Theories About the World Are So Often Wrong*. New York: Basic Books, 2017.

Shtulman, Andrew, and Kelsey Harrington. "Tensions Between Science and Intuition Across the Lifespan." *Topics in Cognitive Science* 8 (2016): 118–37.

Shtulman, Andrew, and Joshua Valcarcel. "Scientific Knowledge Suppresses but Does Not Supplant Earlier Intuitions." *Cognition* 124 (2012): 209–15.

Sinatra, Gale M., and Suzanne H. Broughton. "Bridging Reading Comprehension and Conceptual Change in Science: The Promise of Refutation Text." *Reading Research Quarterly* 46, no. 4 (2011): 374–93.

Sloman, Steven A., and Phillip Fernbach. *The Knowledge Illusion: Why We Never Think Alone*. New York: Riverhead, 2017.

Slovic, Paul, Melissa L. Finucane, Ellen Peters, and Donald G. MacGregor. "The Affect Heuristic." *European Journal of Operational Research* 177, no. 3 (2007): 1333–52.

Stanovich, Keith E. "Rationality, Intelligence, and Levels of Analysis in Cognitive Science." In *Why Smart People Can Be So Stupid*, edited by R. J. Sternberg, 124–58. New Haven: Yale University Press, 2002.

Stanovich, Keith E., and Richard F. West. "Reasoning Independently of Prior Belief and Individual Differences in Actively Open-Minded Thinking." *Journal of Educational Psychology* 89 (1997): 342–57.

Tippet, Krista. "Interview with Daniel Kahneman." *On Being*, October 7, 2017.

White, Peter A. "The Impetus Theory of Judgments About Object Motion: A New Perspective." *Psychonomic Bulletin and Review* 19 (2012): 1007–28.

5

How Do Individuals Think
About Knowledge and Knowing?

As a budding science major, Alex is excited about taking Intro to Biology during his first semester in college. His close-knit family and church community in the remote, rural part of the state are proud of how well he has done in high school, enough to earn him a scholarship and a place in the honors program at the flagship university, albeit a long way from home. At his graduation party Alex gets teased and warned about the ideas he is likely to encounter and admonished to "remember where you came from." He's not concerned as his connection to his family's values feels deep and enduring.

Arriving at college, Alex is not surprised to find that not everyone thinks as he does or that not everyone is religious. He admires the professors who teach his courses, the depth of their knowledge, the clarity of their presentation. Inspired by them, he sees himself becoming a research scientist or a doctor. Then he finds himself in the Intro Bio lectures on Darwin, evolution, and the development of antibiotic-resistant bacteria. The basic principles of evolution appear to be deeply contradictory to what he has learned and what his pastor says the Bible teaches, that God created all living beings, including humans, in their current form.

Studying for the first exam and eager to do well in this course, he reviews the textbook carefully, goes over his notes, and prepares how to answer the questions in ways that align with what he knows the professor will accept as correct, even if he doesn't believe it himself. He begins to struggle with what he thinks is true, trying to reconcile what seem to be competing authorities of religion and science, and of family and science educators.

Interviewed by Barbara in mid-November, Alex flushes and stammers when asked what he thinks about evolution. He reports that he now understands the scientific view of evolution yet is nervous about the implications of abandoning the creationist beliefs taught by his church. He talks about his fears of returning home for Thanksgiving, seeing his pastor and his family for the first time. "If I believe that, who will I be?" he says poignantly. He worries they

Science Denial. Gale M. Sinatra and Barbara K. Hofer, Oxford University Press. © Oxford University Press 2021.
DOI: 10.1093/oso/9780190944681.003.0005

will be appalled by what he is now accepting as scientific theory, confirmed by
an abundance of evidence. He also knows how absurd it would be as a doctor
to not accept that bacteria have evolved to become drug-resistant. It will take
him time to understand that many people with strong religious faith also accept
scientific facts, some at odds with literal interpretations of biblical teachings.
To accept and incorporate what he has learned, he will likely move beyond the
absolutist, dualist thinking that guided him in high school and make room for
nuance, for an evidence-based approach to science, and for a more complex
worldview about knowledge and knowing.

In everyday encounters with new information, conflicting ideas, purported
facts, and assertions made by others, we each have to decide who to believe
and what to accept as true. When we search online about a personal health
issue or how to solve a parenting dilemma or the pros and cons of an issue at
the ballot box, we draw on what others claim to know. Whether we are always
aware of it or not, we each hold beliefs about what knowledge is, what we accept as credible and valid, who we trust as a reliable source of information,
and how we think we know. We also have beliefs about science knowledge
specifically.

Can I trust what scientists tell me? What's the best source of information?
Do I understand what I am reading, or can I just accept that it is true because
of where I read it or who said it? What if my friends or family members have
ideas that conflict with those of experts? What if experts disagree with each
other? These are questions that involve thinking and reasoning about knowledge and knowing, or what psychologists call "epistemic cognition."[1] These
processes help explain issues of science denial, doubt, and skepticism and are
one basis for addressing these issues in our educational systems and in public
discourse.

"Epistemic" comes from the Greek word for knowledge, *episteme*, and
"cognition" refers to thinking and other mental processes. We engage epistemic cognition when we work to resolve competing claims about what
is true about a topic, justify what we know, or evaluate a source of knowledge. Epistemic cognition also encompasses an understanding of what various disciplines—history, science, math, etc.—count as knowledge and how
knowledge is developed within a field.[2] For example, individuals need to understand the basic premises of science as a means of knowing, what educator
Josh Beach describes as "how knowledge is constructed, evaluated, debated,

disseminated, and used" by scientists.[3] Not understanding these assumptions about science can impede acceptance of scientific knowledge.

Epistemic cognition matters as these processes influence how students learn as well as how individuals search for information online, evaluate what they read on social media, interpret and apply scientific knowledge, and engage in discussion and disagreements. It also matters at a collective, societal level. An educated citizenry depends on individual competence to assess conflicting ideas, employ evidence, and develop coherent arguments. Educators have the ability to influence this aspect of cognitive development and to help students develop such productive, adaptive, flexible thinking.

The Development of Epistemic Cognition

Individuals develop their ideas about knowledge and knowing—their epistemic cognition—over time and in patterned ways. A simple way to understand the progression is through a developmental model of three progressive worldviews.[4] Imagine this late-night conversation in a college dorm, where a student studying for an exam with others in his class raises the topic of evolution:

JAY: I just can't buy what we're being taught about evolution, and I know it's not true because my pastor, my parents, and the Bible all say that humans were created in their own form, and I have always believed what they say. What I'm hearing in class has to be wrong.
RICK: But Jay, people are entitled to their own opinions! Some people believe in evolution and some people don't. Everyone has the right to decide.
EMILY: Guys, there's scientific evidence for evolution! This isn't just some wild idea, it's something scientists have studied carefully, and the theory is supported. Let me explain how we know this.

Exemplified in this simplistic dialogue are three discrete ways of knowing that researchers have identified as characterizing a developmental progression: *absolutism* (Jay), or thinking in absolute terms; *multiplism* (Rick), entertaining multiple points of view as equally valid; and *evaluativism* (Emily), evaluating and seeking to substantiate one's views.[5]

Those whose ways of knowing are absolutist are likely to think in black and white, dualistic, dichotomous terms. Knowledge is viewed as unquestioningly objective, certain, and true, handed down intact from authority. Students who view knowledge in this way are likely to see it as a collection of facts, to expect teachers and other authorities to know everything worth knowing, and to view learning as the task of memorizing and repeating back.[6] There are no gray areas and no role for interpretation or subjectivity. For some individuals this dualistic thinking prevails into adulthood. They seek certainty, trust authorities absolutely, and dismiss ambiguity. Science deniers, for example, have been described as engaging in dichotomous thinking, imagining that either facts are known with certainty or there is simply inconclusive controversy.[7] They may rely on known authorities without evaluating their expertise or expecting evidence as a basis for their claims. They may be averse to ambiguity and locked in the stage of absolutism that many outgrow in adolescence. Contemporary examples abound.

Absolutist thinking starts to crumble when individuals recognize that there might be different perspectives on the same topic, that knowledge is not just handed down intact but is constructed by the knower, and that not everything can be known with complete certainty. This can usher in a worldview of *multiplism*, a form of radical uncertainty, where knowledge is subjective, based on interpretation and opinion. All opinions seem equally valid, and each person is entitled to think what they wish, with no means to evaluate conflicting ideas. In Barbara's studies of college students' thinking about evolution, for example, those who held multiplist views were less likely to accept the scientific consensus on evolution than were those who thought evaluatively, the stage that follows. They were also more likely to endorse teaching both evolution and intelligent design (the untestable idea that an intelligent creator guided the development of life) in science classrooms. The language they use for such endorsement is often one of "belief," as exemplified in this student's rationale: "If there is more than one theory, both should be taught. It is better to present all the information available and let the students make their own decisions about what to believe."[8]

The third stage is *evaluativism*, a mode of thinking achieved by a much smaller number of people. Learning to coordinate objectivity and subjectivity, individuals acknowledge that there may be multiple perspectives on various issues, but some can be ascertained as more valid than others. Knowledge is viewed as contingent and contextual, and there are acknowledged criteria for evaluating authorities and assessing competing ideas about what is presented as factual. Evidence plays a significant role in determinations of what to accept as true, a habit of mind that enhances science understanding. As 16-year-old climate activist Greta Thunberg posted on Twitter, following the 2019 climate strikes, "To those who question my so-called 'opinions,' I would once again want to refer to page 108, chapter 2 in the SR1.5 IPCC report released last year. There you'll find our rapidly declining CO_2 budgets. This is not opinion or politics. It's the current best available science." At this level, expertise can also be evaluated, rather than accepted unilaterally. Individuals seek a warranted trust in expertise. Progression to this level is related to educational experiences and can be nurtured in both formal and informal learning contexts.[9] Within these three worldviews are evolving conceptions of the nature of objectivity and subjectivity, the certainty and simplicity of knowledge, the source of knowledge, and justifications for knowing,[10] transformed and reconfigured at each level.

Although cognitive development and education are critical for the higher levels of such competence, societal influences may also affect individuals' worldviews about knowledge. The current "post-truth" era[11] can be a haven for those who think in absolute terms, allowing blind rejection of scientific facts, such as climate change, because their preferred authorities tell them to, in spite of all evidence to the contrary. Absolutist thinkers also have been given new criteria for a dichotomous view of reality: "Facts are no longer correct or incorrect. Everything is true unless it's disagreeable, in which case it's fake."[12]

Post-truth messaging primarily fosters a view of knowledge that is multiplistic, however, in which individuals imagine that any claims about what is true are only opinions anyway, so how could we really know? When facts become subordinate to emotion, personal belief, and political point of view,[13] multiplistic thinking is fostered. Tom Nichols in *The Death of Expertise* provides a clear description of the problem: "It is a new Declaration of Independence: No longer do we hold *these* truths to be self-evident, we hold *all* truths to be self-evident, even the ones that aren't true. All things are knowable and every opinion on any subject is as good as any other."[14]

When accepting multiple points of view becomes entrenched as a worldview, it can be claimed as an aspect of individual rights, each with entitlement to their own beliefs, however unsubstantiated. Sometimes this stance involves a confusion of relativism and tolerance. In Barbara's research she has interviewed students who in well-meaning ways seem to have overgeneralized allowing for a diversity of others' beliefs, extending this acceptance to ignorance about issues for which there is strong empirical support. What is perceived as a core skill set in higher education—argumentation, refutation, and the marshalling of significant evidence to support contradictory claims—may be avoided by students who imagine open disagreement as offensive to others. Extreme relativism can lead to intellectual nihilism, a pointlessness about confirming truth or fact, a sense that truth itself does not exist, so why bother with effortful vetting of information? Unfortunately, this can play into the hands of those who seek to discredit scientific findings, whether about climate change, the safety of vaccinations, or any other such topic. As Daniel Moynihan has been widely quoted, "Everyone is entitled to his own opinions, but not to his own facts."

Fueling Multiplistic Thinking Through False Equivalencies: Balance as Bias

Multiplism can be fostered by a press too eager to present "both sides" of issues that are not actually controversies in the scientific community, such as the human role in climate change. When the media seek to present "balanced" coverage on topics already adjudicated by scientific research, this false equivalency prevents citizens from understanding the facts and evidence that do exist. Although journalists may want to create controversy as a means of building readership or as a rhetorical style, creating tension in the story, this is an intellectually irresponsible practice on topics where scientific consensus does exist, and a "balanced" approach can falsely cloud public understanding and judgment. Awareness of the problem is growing, following an independent review of BBC science coverage published in 2011, which concluded that the network's practices of creating false balance made scientific debates on topics such as climate change, genetically modified organisms (GMOs), and vaccinations appear more controversial than they are.[15] Follow-up studies were conducted in 2012 and 2014, and in 2018 the BBC sent a briefing note to all staff, acknowledging that human-made climate change exists and should be reported and alerting them to be wary of false balance. "To achieve impartiality, you do not need to include outright deniers of climate change in BBC coverage, in the same way you would not have someone denying that Manchester United won 2–0 last Saturday. The referee has spoken" (https://www.theguardian.com/environment/2018/sep/07/bbc-we-get-climate-change-coverage-wrong-too-often). News organization such as the BBC have also recommended "weight of evidence" reporting, indicating where the predominant expert opinion lies, although this has been criticized as allowing room for the outlier's take on the issue.

Yet the problem of false equivalency persists, with journalists eager to portray controversy even on scientific issues that have long been decided and about which there is no scientific debate whatsoever. When Barbara was interviewed by a WHYY reporter assembling a story on flat Earth believers for *The Pulse*, a science podcast, she was taken aback by the reporter's portrayal of "two sides" of the issue, described as "flat Earth believers and round Earth believers." Barbara responded that these were not both "beliefs" but that one was an erroneous and unsubstantiated belief and the other was accepted scientific knowledge. To portray them as commensurate was a disservice to the public, a comment that was not well received by the interviewer. Journalists may be so accustomed to presenting "both sides" as a goal of fairness (or are perhaps eager to stir up readership) that they can become complicit in the representation of science as a set of competing beliefs—even when the concept is as settled as whether Earth is flat or round.

Balance can actually *become* bias, when it influences popular discourse to diverge from scientific discourse. An analysis of global warming coverage by the US "prestige press" (the *New York Times*, the *Los Angeles Times*, the *Washington Post*, and the *Wall Street Journal*) from 1988 to 2002 showed just how strong such informational bias can be, particularly in regard to a focus on uncertainty about the issue, inconsistent with prevailing scientific wisdom.[16] In some cases, doubt is manufactured in the minds of the public by firms that exploit and exaggerate scientific uncertainty,[17] through the hiring and funding of scientists who participate in science-denying communication that favors particular industries.[18] Journalists contribute to the social construction of ignorance when they report industry claims designed to sow public confusion.[19]

The results of reporting false equivalencies and giving disproportionate visibility to outlier perspectives are not insignificant. Experimental studies by psychologist Derek Koehler show how presentations that attempt to show "balance" on issues where experts agree can confuse and distort perceptions of expert judgments.[20] In one study participants were given numerical summaries of expert opinion on economic issues, including ones where a large majority of expert economists agreed, such as whether a carbon tax would be an efficient means for reducing carbon dioxide emissions. One group of participants saw only the numerical summary, which showed that 93 experts agreed, 5 were uncertain, and only 2 disagreed. Another group not only saw the summary but also read comments by an expert on each side of the issue, one selected from the 93 who agreed and one from the 2 who did not. This is a ratio similar to that of climate scientists' acceptance of climate change, and many news outlets have pitted one supporter against one denier, giving deniers proportionally more media coverage than the 98% of concerned climate scientists.[21]

Why does this matter? In Koehler's study the competing expert comments diminished the degree to which participants perceived consensus, even though both groups had also seen the numerical data. Reading the comments also influenced their judgment about whether consensus was sufficient to guide public policy. In other words, seeing conflicting testimonies, even when the accompanying data were overwhelmingly conclusive, weakened acceptance of strongly supported expert judgments. In a commentary on the study, Koehler suggested that such presentations give us mental representations of a person on each side, distorting our sense of how representative each side is, and that it may be hard to discount a plausible argument from an expert, even if it is known to be held only by a minority.[22] Moreover, hearing dissenting expert opinion may induce a sense of uncertainty, even about topics where experts have reached firm conclusions and only a small minority dissent.

Educators may also contribute to such misunderstandings and undermine the acceptance of expert knowledge. Imagine the student who needs to identify information on a range of current socio-scientific topics such as the safety of GMOs, or envision the individual seeking what they hope will be unbiased information on a meaningful topic of interest, perhaps vaccine safety. Both are likely to come upon ProCon.org, a web site that serves more than 20 million people annually, including teachers and students in all 50 states and 90 countries. So what does a pro–con format convey? Well, imagine a student who wanted to research any of the following: "Do violent video games contribute to youth violence?" "Should any vaccines be required for children?" "Is human activity primarily responsible for global climate change?" "Should genetically modified organisms be grown?" "Is cell phone radiation safe?" The site is a high school debater's dream; no matter which side you happen to be assigned, there is a roughly equivalent number of pros and cons on each of these topics. For those who do not yet think evaluatively and who would be likely to click on the footnotes, investigate the sources, and go beyond this website, they might learn to see any of these issues—and the dozens of others that are profiled on the site—as controversial and unsettled, simply a host of opinions, without expert consensus. This is a significant disservice to students, teachers, and other adults making use of the site. Such approaches help explain both the durability of multiplistic thinking and the dire need to foster critical epistemic thinking. This particular site, and others like it, would benefit from the BBC's internal dictate to show where the preponderance of evidence lies, albeit with the awareness that the false equivalencies of balancing pros and cons on settled topics such as climate change can still weaken acceptance of expert judgments, as Koehler's research shows. Teachers also need to question any press to "teach the controversy" in situations where the preponderance of evidence suggests there is no actual controversy, such as the theory of evolution.

Scientific Literacy and the Practice of Science

What is science, how is it conducted, and what basic principles guide scientists' endeavors? How do scientists develop and verify knowledge? The degree of public understanding of scientific claims can often be linked to conceptions—and misconceptions—of the scientific enterprise itself. Science literacy involves more than knowing about disciplinary core ideas or simply learning scientific facts and findings; it also involves an understanding of the epistemology of science, how it is produced and validated.[23] In other words, students need to learn not only *what* scientists know but *how* they know.[24] This has been advanced by the Next Generation Science Standards (NGSS),[25] a collaborative effort among the National Research Council, the National Science Teachers Association, the American Association for the Advancement of Science, and others to create research-based science education standards for implementation by teachers at the local level. These standards include the content knowledge that teachers are expected to address as well as the expectation that K–12 students will understand the key characteristics of the scientific enterprise, the practice of science, and science as a way of knowing.

These national science education standards, in plain-speaking language, describe four basic premises shared by scientists, the fundamental beliefs and attitudes that guide their practices: 1) the world is understandable through systematic study; 2) science is a process for discovering knowledge, and scientific ideas are subject to change; 3) most scientific knowledge is durable, even as scientists reject the notion of absolute truth and continue to modify scientific ideas; and 4) science cannot provide complete answers to all questions as some matters cannot be examined in a scientific way.[26]

Fundamentally, scientific inquiry depends on evidence.[27] Although methods of inquiry vary widely, well beyond the "scientific method" of hypothesis testing and controlled experimentation taught in most schools, scientists share an understanding that the validity of their findings depends on their observations and the empirical evidence they provide. Many news readers were repeatedly reminded of this need during the early months of the coronavirus pandemic, as ideas about treatment for COVID-19, the disease caused by the virus, were spread by political leaders and others. Hydroxychloroquine, a drug used to treat malaria and lupus, was touted as a possible "game changer," but it had limited testing for treatment of COVID-19 and potentially serious side effects, suggesting the need for further study before any recommendations could be made. Such treatments need to be based not on anecdotal evidence, intuition, or hunches but on empirical evidence, carefully validated.

In addition, as historian of science Naomi Oreskes notes, science is consensus-based,[28] dependent on peer review of expert colleagues who vet the knowledge developed, a process she and her colleague Erik Conway describe as "what makes science science."[29] Scientific claims require the critical scrutiny of multiple experts to achieve legitimacy. This process, however, takes time, and as the pandemic raged, many became impatient with the process. In June 2020 the *New York Times* warned that thousands of papers were being rushed to online sites with little or no peer review, given the pressure to find treatments.[30]

Misconceptions About Science as a Way of Knowing

In spite of the simplicity of these basic tenets of science, they may be overlooked in science teaching, which is often overly focused on the content knowledge addressed in standardized testing. Accordingly, the public may have erroneous views of science as a way of knowing. Such misconceptions can undermine acceptance of scientific knowledge. A sampling of these epistemic confusions follows:

"But it's just a theory." One of the misconceptions that prevents students from accepting the theory of evolution, for example, is a confusion of "theory" with "hypothesis". They typically apply the colloquial use of theory ("You lost your keys again? I've got a theory about that!") as a general hunch about how something might function. The risk is that they will dismiss substantial scientific findings as "just a theory."[31] In Barbara's research on student understanding and acceptance of evolution, she and her colleagues asked college students to define "theory" in their own words. In the context of a survey where the headings and questions referred to the "theory of evolution," students' descriptions of the term showed that most assumed it was an idea or a hunch, a hypothesis, or a partially proven or unproven idea that was often widely accepted.[32] Only 7% of the college students surveyed gave a definition consistent with scientific thinking, viewing it as a broad, well-substantiated explanation for some aspect of the natural world, based on tested hypotheses, with predictive power. Students were also asked about whether they favored the teaching of intelligent design in schools, and they gave answers consistent with such misconceptions: "Evolution is still strictly a theory, so there is no reason for it to be the only one taught." "If there is more than one theory, both should be taught." "The theory of evolution is not any more proved than the idea of intelligent design." Students need help in disentangling these misconceptions, beginning with learning how scientists use terms that are fundamental to the field and how that might differ from common use of the same terms.

Uncertainty in science. Theories, however, are open to revision, as is all scientific knowledge. Philosopher Lee McIntyre sums up what he calls "the scientific attitude" as a commitment to two principles: caring about empirical evidence and a willingness to change theories in light of new evidence.[33] One of the fundamental tenets of science is that knowledge is open to revision, yet this is the crux of frequent misunderstandings in the public confusion about science. By contrast to what scientists view as the ever-evolving nature of scientific knowledge, the average layperson may assume that what scientists know is certain and represents both proof and complete consensus. When that is not the case, they may be skeptical, frustrated, doubtful, and dismissive. With the rapid spread of coronavirus, for example, scientists and medical experts could not be expected to have all the information needed at the outset; and as more evidence became available, advice sometimes changed. For those who expect certainty, this might be seen not as progress but as a sign of failure, potentially breeding distrust.

How do individuals interpret the statistic that 98% of climate scientists agree the causes of global climate change are human-induced? Some may see this as well more than adequate confirmation; others may interpret it to mean that no certain conclusion has been achieved, so it is still debatable. Scientists qualify their findings, hedging with terms such as "indicates that" or "provides support for the hypothesis" and seldom assert that their findings *prove* a claim. This commitment that research is ongoing is critical to the profession. Although some scientific facts are clearly established (e.g., the law of gravity), some research results simply cannot be certain, such as forecasting the timing and impact of climate change. As Norwegian psychologist and economist Per Espen Stoknes notes, projections of future climate change scenarios are, of course, uncertain, and "science always deals with estimates and probabilities, not with absolutes."[34] Yet "fear of uncertainty" in science can drive individual decision-making. Researchers who studied parent conversations on Facebook after the 2013 polio crisis in Israel have reported such fear of uncertainty as a basis for some vaccine hesitancy.[35] Parents would understandably prefer 100% certainty about efficacy or lack of side effects, but such guarantees are unlikely.

The flip side of this issue is when journalists report on scientific findings and drop hedging language altogether, giving the impression of conclusiveness before it is warranted. This type of reporting might make for a cleaner story and snappy headlines ("Chocolate eases depression!"), but it is misleading. This presentation of certitude is exacerbated by the fact that people often prefer "false clarity" to the task of processing nuance. Such a preference can impair the ability to detect the tentative nature of findings in health news, pending further confirmation, even when it is presented as part of the story.[36]

"The scientific method." Overschooled in a narrow view of the scientific method that involves controlled studies, some individuals think that science is conducted only in labs under experimental conditions. This may lead them to be dismissive of findings from areas of science that involve analysis of data from scientific observations. They may misunderstand the processes scientists use to learn about evolution, climate change, geological ages, planetary motion. They may question the knowledge base of those who are extrapolating from current evidence, to understand either past events, such as how the dinosaurs became extinct, or forecasted ones, such as the rise in sea levels. In another of Barbara's studies of students' epistemic thinking, students from 6th to 12th grade were observed thinking aloud as

they searched online to learn how bees communicate. Many reasoned concretely and were suspicious of processes of scientific inference. As one student noted, it would be impossible to know how bees communicate "unless you had a bee decoder." Asked also about the demotion of Pluto from planetary status, some were skeptical that such a conclusion could be reached as scientists had not actually been to Pluto. As one student succinctly phrased it, "You can't know unless you go." With belief that science is a hands-on observable activity, students had trouble seeing how scientists could make distinctions that seemed arbitrary, abstract, and inferential. Adults who cannot distinguish between the weather they personally experience and scientists' claims about changing climate may be falling into similar traps. Flat-earthers also privilege the concrete experience of their own lives. As reported in the *New Yorker*'s coverage of a flat-earth conference, a presenter suggested that "99% of received wisdom is questionable; if you can't observe it for yourself, it can't be trusted."[37] This type of thinking is particularly problematic for those who have been unable to accept the serious harm possible from coronavirus, an invisible agent, and thus dismissed it as a hoax or thought COVID-19 was similar to a typical flu—until someone they know contracted it.

Epistemic Trust in Science

Individuals simply cannot have adequate knowledge on many complex scientific topics. They may not have the interest to care about understanding some issues or the time and background knowledge to get up to speed on those that do matter to them. In any of these common situations, individuals are likely to rely on the judgments of experts, those in the scientific community with more thorough knowledge. Relying on the expertise of others, whether a doctor, a professor, a news source, a medical journal, or other such authorities, involves *epistemic trust*. We choose who we respect as reliable sources of the knowledge that we do not have ourselves but need in order to make decisions about our health, our communities, our planet. Such trust is fundamental to the public understanding of science. The importance of this trust has seldom been more evident than in the spread of coronavirus, as individuals were left to sort out whose expertise to trust, given often conflicting advice from political leaders, the Centers for Disease Control and Prevention, the World Health Organization, state health officials, and family doctors. In May and June 2020 many states began to allow businesses to re-open, in direct contrast to data that showed the virus had not waned. Those who trusted their governors or local leaders to make such decisions may not have been aware that the openings were guided more by economic interests than medical expertise, with catastrophic results for many regions of the United States.[38]

The public has varying degrees of trust in the scientific community as a whole, as well as trust of particular scientific authorities.[39] An August 2019 study by the Pew Research Center indicates that most Americans have an overall confidence that scientists act in the public interest, with 86% of those surveyed expressing either "a fair amount" or "a great deal of confidence."[40] Those with stronger science knowledge exhibited greater confidence than those with less factual science knowledge. Confidence in medical science parallels that of scientists overall (87%), but trust in nutritional science researchers was considerably lower, with only 51% expressing a mostly positive view. Trust in scientific and medical authorities to influence policy, however, is complicated by the degree to which "values" influence such trust. For example, a May 2020 survey by *Scientific American* found that whether individuals believed that experts (such as Anthony Fauci, director of the National Institute of Allergies and Infectious Diseases) should guide policy depended on whether they thought such experts shared their values, with conservatives showing more skepticism.[41]

Such issues of trust play out regularly in the media. Not long after the Pew survey results were published, a *New York Times* headline, on September 30, 2019, proclaimed, "Eat Less Meat, Scientists Said. Now Some Believe That Was Bad Advice." And in bold print under the headline, this summary of expert backlash: "The evidence is too weak to justify telling individuals to eat less beef and pork, according to new research. The findings 'erode public trust,' critics said." Perhaps you, too, were skeptical when you read what seemed to contradict long-term medical advice to eat less red meat, in-cluding a report the New York Times ran during the same month: "Getting Most of Your Protein from Plants May Help You Live Longer." What is a concerned consumer to do with these stories? Notably, the Pew survey also identified a lack of transparency about conflicts of interest with industry as a key problem in trust, with only 12% of respondents agreeing that nu-trition researchers display such transparency and only 11% agreeing that researchers take responsibility for their mistakes.[42] Within a week after, the *Times* sheepishly ran another story, providing an exemplar of the Pew findings on distrust: "Scientist Who Discredited Meat Guidelines Didn't Report Food Industry Ties." Journalists need to apply more scrutiny before leaping to report, or they foster undermining trust.

Erosion of epistemic trust also happens when peer review fails, as when vaccinations were falsely linked to autism, which led to the retraction of the paper by *The Lancet* (albeit many years later) as well as revocation of the author's medical license.[43] Yet the beliefs espoused in the original piece persist, leading to loss of confidence in the vaccine that prevents measles, mumps, and rubella, and thus to serious outbreaks and multiple deaths. Others have become suspicious of the authority of medical journals, given the failure of the review process to prevent such a study from publication in a prestigious journal. More recently, a study of the effectiveness of chloroquine and hydroxychloroquine in treating COVID-19 was retracted from the same journal when the authors were unable to complete an independent audit of their data and said that they could "no longer vouch for the veracity of the primary data sources."[44]

With devastating ramifications, the claim that opioids for pain treatment were unlikely to be addictive, promulgated by pharmaceutical firms and accepted by doctors and then believed by their patients, proved to be indubitably false. Such a chain of unearned trust has led to a major public health crisis and an astonishing half a million deaths in the United States by 2018, a number that was continuing to grow at the rate of 130 deaths per day, according to the Centers for Disease Control and Prevention.[45] Many now wonder what to do for pain management and whose recommendations are reliable. They may be particularly and understandably suspicious of profit-motivated pharmaceutical companies and those doctors who relied on them, some of whom appear to have been swayed by their own personal gain.

Epistemic trust of science may be especially compromised in marginalized communities, and the reasons are often complex. African Americans who know of the Tuskegee Syphilis Study have long been assumed to view medical research science as an institution unworthy of their trust, suspicious of how minorities are utilized as research participants. Conducted between 1932 and 1972 by the US Public Health Service, and notoriously titled the Tuskegee Study of Untreated Syphilis in the Negro Male, the study followed 600 Alabama sharecroppers under the guise of treatment for bad blood, although the men were never treated—not even after penicillin was determined to be a successful treatment in 1947. The official protests of a Public Health Service whistleblower in 1966 and 1968 were dismissed, and only when he leaked the information to a reporter did the concern rise to the level of governmental attention. Notably, however, the Tuskegee Legacy Project study found no actual differences between Blacks and Whites in more recent willingness to participate in medical research.[46] Assuming that differences in such willingness persist could become a scapegoat for soliciting low minority participation. Moreover, applying assumptions about Tuskegee to vaccine hesitancy among minorities during the COVID-19 crisis may mask issues of trust that arise from unequal access to medical care, for example.[47]

Women were historically excluded from many clinical studies, although the results were presumed to be generalizable to them. Only in 2001 did the Institute of Medicine conclude that sex should be recognized as an important variable in research.[48]

Trust is critical both in doing science and in the public understanding of science.[49] Scientists rely on knowledge produced by other experts, and they participate in a peer-review process that allows peer experts to function as gatekeepers for what counts as the state of knowledge about a topic—always subject to continued revision, of course. What the public has every right to expect, according to philosopher and epistemologist Heidi Grasswick, is that the scientific community serves as a filtration system, "sorting through the wealth of research available and scientific knowledge being produced, finding for us both the sound and the most significant."[50]

Because laypeople have only a bounded understanding of science yet access to a wealth of information online, they must make decisions about what and whom to trust.[51] Science communicators, whether the scientists themselves or those in a position to convey science to the public, need to convey trustworthiness as well as expertise. Frederike Hendriks and her colleagues at the University of Muenster have examined how people decide whose help is dependable when they seek scientific information from experts online. They found that laypersons are less likely to ask themselves *Is this true?* than to ponder *Whom can I believe?* Their multiple empirical studies show that individuals answer the latter question by considering expertise, integrity, and benevolence.[52]

What Can We Do?

Each individual has beliefs about knowledge and knowing, as well as about how science is conducted. These ideas may influence understanding of scientific claims and affect learning, reasoning and thinking about science. Each of us can work to improve our own epistemic competence, as well as help facilitate this process for those we communicate with or teach.

What Can Individuals Do?

First, value the search for truth. Prize it, honor it, respect it, seek it. Sharpen your mental arsenal, and call out unsubstantiated claims. Be wary of attempts to discredit scientific knowledge that work in favor of the messenger. Tobacco companies sowed doubt about the health risks of smoking, Exxon buried and then refuted knowledge of climate change and now promotes solutions at the individual and not corporate level, evading responsibility. These "merchants of doubt,"[53] those with financial gain to be made at the expense of undermining science, are also rivaled in the political sphere. Be alert to attempts to cast known truths as unsettled ones. Don't allow the alarms about a "post-truth" society to promote a surrendering of the value of truth. Concomitantly, know that scientific findings continue to accumulate, are open to revision, and are refined over time.

Marshall epistemic vigilance. Take the time to question, evaluate, and monitor the information you are seeking and receiving, whether in online searches, in social media, in the news, or in conversations. Vigilance is the opposite of blind trust and requires astute attention. We cannot be vigilant all the time, nor do we need to be; we have established our own shortcuts to know when trust is likely warranted and when it might not be. Review those heuristics regularly, however. Are you being well served by the sources you trust? Diversify and triangulate. Note when you employ vigilance and why. Strengthen the muscle. It is easy to get intellectually lazy and not inquire at a deeper level about sloppy conclusions, misrepresentation of data, or inherent biases.

Value scientific evidence. At its core, science is an evidence-based practice. Question suspicious claims, ask for support, review the evidence. If you are handed a book by a friend that is purported to argue for a "slow vax" approach to inoculations, check it out thoroughly. One that a colleague enthusiastically recommended to Barbara was written by a doctor whose license had been revoked, and it provided no scientific evidence to support this approach. Seek peer-reviewed scientific evidence, not just anecdotes or spurious correlations.

Employ your science knowledge, and continue to educate yourself about science. Understand the diversity of scientific methods. Recognize that a tentative conclusion is not an invalid one. Appreciate clinical studies designed with experimental controls (where feasible). Take the time to interpret the graphs that accompany what you read. Assess the credentials of authors, and seek peer-reviewed material.

Listen to those with whom you do not agree on scientific issues. Listen to others, and find out if they value truth, know how to seek it, and have evidence to support their statements. Perhaps you can recognize their epistemic perspectives and see if you can discern when someone might be operating from a dualistic or multiplistic perspective. Help others recognize the value in evaluating claims about what is true and learning how to discern what gets accepted and why. Keep an open mind, and learn to examine your own epistemic perspectives as well.

What Can Educators Do?

Teach how scientific knowledge is produced and what the underlying premises are. Review the NGSS for more information. Yes, teach "the scientific method" so that students understand hypothesis testing and experimental designs, but go beyond it to explain how science is conducted in different fields. Help students understand how scientific knowledge is produced and validated, what peer review is and why it matters, and why scientists are cautious in overstating their research findings.

Teach the value of scientific evidence. Help students understand that evidence is the basis of scientific endeavors. Help them know what counts as evidence and how it accumulates and why it matters. Expect them to provide scientific evidence for their statements during class discussions, and teach them how to evaluate what they find.

Provide training in the evaluation of scientific expertise and scientific claims. Give students monitored practice in online searching about scientific ideas so that they learn real-world tasks of searching and evaluation. These are likely some of the most important skills they can learn in order to continue learning throughout life. Foster digital literacy, and work with your library staff to create means of teaching it that will be genuinely useful to students.

Foster epistemic vigilance. Help students develop habits of mind that keep them alert to fraudulent statements and representations. Assist them in discerning the motives of those who wish to persuade. Teach them why it matters. Acknowledge their progress in turning on and developing this skill set.

Nurture evaluativist thinking and epistemic competence. Consider students' epistemic perspectives when they marshal arguments in class or in papers, and meet them where they are. The advantages of an evaluativist perspective are considerable, but it takes time to get there. Learners will need to alter their beliefs to accept that scientific assertions are a result of evaluated judgments, not opinions.[54] Promote epistemic competence in ways that extend beyond the classroom, a significant challenge.[55] Teach for transfer, giving students opportunities to use what they are learning in their own search for knowledge about topics that interest them.

Teach and model epistemic virtues. Educational philosopher Jason Baehr advocates a set of virtues for education: open-mindedness, curiosity, intellectual courage, humility.[56] These are the virtues that undergird the practice of science as well. Teachers can strive to model these virtues and to acknowledge them in their students.

What Can Science Communicators Do?

Include information about the premises of science. Convey the content of a scientific topic, but also remind readers of the underlying premises of the scientific enterprise when it can help with interpreting the story. How do scientists know what they know? Help readers understand the diversity of methods that underpin various fields.

Provide evidence for scientific claims. Give readers and viewers enough information to evaluate research findings more fully. How were the conclusions reached? Show logical chains of reasoning, especially helpful in online websites that get frequent use. As an example, NASA has a simple presentation of "What is climate change?" with four buttons to click, simply identified as Evidence, Causes, Effects, Solutions.[57]

Avoid creating false equivalencies in the guise of "fair and balanced" or presenting "both sides" when the science is in fact settled. It does a disservice to the public to give equal time to a climate denier and a climate scientist in a piece about climate change, leaving viewers to conclude that the findings are ambiguous and logically contended. We do need to know more about how anti-vaxxers think, and such interviews can be helpful in developing effective health campaigns. That type of article is different from presenting them as one of two sides in an argument about the effectiveness of vaccinations, countering pediatricians' point of view. Celebrities promoting unsubstantiated views have been given an undue amount of attention on that particular issue.

Conclusions

Epistemic cognition, although it may sound esoteric, lies at the heart of how we each make sense of the massive amount of information available to us— what we value as expertise, how we justify what we know, and how we evaluate others' claims. Knowing more about it may benefit each of us in our own knowledge building, as well as teachers and science communicators.

Notes

1. Barbara K. Hofer, "Epistemic Cognition as a Psychological Construct: Advancements and Challenges," in *Handbook of Epistemic Cognition*, ed. Jeffrey Alan Greene, William A. Sandoval, and Ivar Bråten (New York: Routledge, 2016).
2. Barbara K. Hofer, "Beliefs About Knowledge and Knowing: Integrating Domain Specificity and Domain Generality: A Response to Muis, Bendixen, and Haerle (2006)," *Educational Psychology Review* 18, no. 1 (2006).
3. Josh M. Beach, *How Do You Know? The Epistemological Foundations of 21st Century Literacy* (New York: Routledge, 2018), 6.
4. Deanna Kuhn, Richard Cheney, and Michael Weinstock, "The Development of Epistemological Understanding," *Cognitive Development* 15, no. 3 (2000).
5. Note that this developmental framework is best viewed as a useful heuristic for understanding worldviews about knowledge and knowing, and not as a precise developmental scheme. For a critique, see Sarit Barzilai and Clark A. Chinn, "On the Goals of Epistemic Education: Promoting Apt Epistemic Performance," *Journal of the Learning Sciences* 27 (2018).
6. Barbara K. Hofer, "Personal Epistemology Research: Implications for Learning and Teaching," *Educational Psychology Review* 13 (2001).
7. Jeremy P. Shapiro, "The Thinking Error at the Root of Science Denial," The Conversation, May 8, 2018, https://theconversation.com/the-thinking-error-at-the-root-of-science-denial-96099.

8. Barbara K Hofer, C. F. Lam, and A. DeLisi, "Understanding Evolutionary Theory: The Role of Epistemological Development and Beliefs," in *Epistemology and Science Education: Understanding the Evolution vs. Intelligent Design Controversy*, ed. R. Taylor and M Ferrari (New York: Routledge, 2011).

9. Mark K. Felton and Deanna Kuhn, " 'How Do I Know?' The Epistemological Roots of Critical Thinking,'" *Journal of Museum Education* 32 (2007).

10. Barbara K. Hofer and Paul R. Pintrich, "The Development of Epistemological Theories: Beliefs About Knowledge and Knowing and Their Relation to Learning," *Review of Educational Research* 67, no. 1 (1997).

11. Lee McIntyre, *Post-Truth* (Cambridge, MA: MIT Press, 2018).

12. Alan Burdick, "Looking for Life on a Flat Earth: What a Burgeoning Movement Says About Science, Solace, and How a Theory Becomes Truth," *The New Yorker*, May 30, 2018.

13. McIntyre, *Post-Truth*.

14. Tom Nichols, *The Death of Expertise: The Campaign Against Established Knowledge and Why It Matters* (New York: Oxford University Press, 2017).

15. BBC Trust, "Review of Impartiality and Accuracy of the BBC's Coverage of Science," https://www.bbc.co.uk/bbctrust/our_work/editorial_standards/impartiality/science_impartiality.html.

16. Maxwell T. Boykoff and Jules M. Boykoff, "Balance as Bias: Global Warming and the US Prestige Press," *Global Environmental Change* 14, no. 2 (2004).

17. Yann Bramoullé and Caroline Orset, "Manufacturing Doubt," *Journal of Enviromental Economics and Management* 90 (2018).

18. Naomi Oreskes and Erik M. Conway, *Merchants of Doubt: How a Handful of Scientists Obscured the Truth on Issues from Tobacco Smoke to Global Warming* (New York: Bloomsbury Publishing, 2010).

19. S. Holly Stocking and Lisa W. Holsteing, "Manufacturing Doubt: Journalists' Roles and the Construction of Ignorance in Scientific Controversy," *Public Understanding of Science* 18 (2009).

20. Derek J. Koehler, "Can Journalistic 'False Balance' Distort Public Perception of Consensus in Expert Opinion?" *Journal of Experimental Psychology: Applied* 22 (2016a).

21. Per Espen Stoknes, *What We Think About When We Try Not to Think About Global Warming: Toward a New Psychology of Climate Action* (White River Junction, VT: Chelsea Green Publishing, 2015).

22. Derek J. Koehler, "Why People Are Confused About What Experts Really Think," *New York Times* February 14, 2016b.

23. Gale M. Sinatra and Barbara K. Hofer, "Public Understanding of Science: Policy and Educational Implications," *Policy Insights from the Behavioral and Brain Sciences* 3, no. 2 (2016).

24. Doug Lombardi, "Thinking Scientifically in a Changing World," American Psychological Association, January 2019, https://www.apa.org/science/about/psa/2019/01/changing-world.aspx.

25. National Research Council, *A Framework for K–12 Science Education: Practices, Crosscutting Concepts, and Core Ideas* (Washington, DC: National Academies Press, 2012).

26. NGSS Lead States, *Next Generation Science Standards: For States, by States* (Washington, DC: National Academies Press, 2013).

27. For further explication of the meaning of evidence, see Ravit G. Duncan, Clark A. Chinn, and Sarit Barzilai, "Grasp of Evidence: Problematizing and Expanding Next Generation Science Standards' Conceptualization of Evidence," *Journal of Research in Science Teaching* 55 (2018).

28. Naomi Oreskes, *Why Trust Science?* (Princeton, NJ: Princeton University Press, 2019).

29. Oreskes and Conway, *Merchants of Doubt*, 154.

30. R. C. Rabin and E. Gabler, "Two Huge Covid-19 Studies Are Retracted After Scientists Sound Alarms," *New York Times* June 4, 2020.

31. Lee McIntyre, *The Scientific Attitude: Defending Science from Denial, Fraud, and Pseudoscience* (Cambridge, MA: MIT Press, 2019).

32. Hofer, Lam, and DeLisi, "Understanding Evolutionary Theory."

33. McIntyre, *The Scientific Attitude*.

34. Stoknes, *What We Think About When We Try Not to Think About Global Warming*, 9.

35. Daniela Orr and Ayelet Baram-Tsabari, "Science and Politics in the Polio Vaccination Debate on Facebook: A Mixed-Methods Approach to Public Engagement in a Science-Based Dialogue," *Journal of Microbiology and Biology Education* 19 (2018).

36. Joachim Kimmerle et al., "How Laypeople Understand the Tentativeness of Medical Research News in the Media an Experimental Study on the Perception of Information About Deep Brain Stimulation," *Science Communication* 37, no. 2 (2015).

37. Burdick, "Looking for Life on a Flat Earth," 11.

38. CNN, "19 States See Rise in Covid-19 Cases Amid Reopening and Protests," June 12, 2020.

39. Rainer Bromme and Susan R. Goldman, "The Public's Bounded Understanding of Science," *Educational Psychologist* 49, no. 2 (2014).

40. Cary Funk et al., "Trust and Mistrust in Americans' Views of Scientific Experts," Pew Research Center, August 2, 2019, https://www.pewresearch.org/science/2019/08/02/trust-and-mistrust-in-americans-views-of-scientific-experts/.

41. John H. Evans and Eszter Hargittai, "Why Would Anyone Distrust Anthony Fauci?" *Scientific American*, June 7, 2020, https://blogs.scientificamerican.com/observations/why-would-anyone-distrust-anthony-fauci/.

42. Funk et al., "Trust and Mistrust in Americans' Views of Scientific Experts."

43. Sarah Boseley, "How Disgraced Anti-Vaxxer Andrew Wakefield Was Embraced by Trump's America," *The Guardian*, July 18, 2018.

44. "Retraction: Study on Chloroquine and Hydroxychloroquine in COVID-19 Patients," *The Lancet* (May 22, 2020).

45. Centers for Disease Control and Prevention, "Opioid Overdose: Understanding the Epidemic," https://www.cdc.gov/drugoverdose/epidemic/index.html.

46. Ralph V. Katz, Stefanie L. Russell, and Cristina Claudio. The Tuskeekee Legacy Project: Willingness of Minorities to Participate in Biomedical Research. *Journal of Health Care for the Poor and Underserved* 17(4) (2006): 698–715.

47. "Stop blaming Tuskegee, Critics Say. It's not an 'excuse' for current medical racism." National Public Radio, March 23, 2021.

48. Katherine A. Liu and Natalie A. D. Mager, "Women's Involvement in Clinical Trials: Historical Perspective and Future Implications," *Pharmacy Practice* 14 (2016).

49. Friederike Hendriks, Dorothe Kienhues, and Rainer Bromme, "Trust in Science and the Science of Trust," in *Trust and Communication in a Digitized World*, ed. Bernd Blöbaum (Cham, Switzerland: Springer, 2016).

50. Heidi E Grasswick, "Scientific and Lay Communities: Earning Epistemic Trust through Knowledge Sharing," *Synthese* 177, no. 3 (2010).

51. Bromme and Goldman, "The Public's Bounded Understanding of Science."

52. Hendriks, Kienhues, and Bromme, "Trust in Science and the Science of Trust."

53. Oreskes and Conway, *Merchants of Doubt*.

54. Deanna Kuhn and Seung-Ho Park, "Epistemological Understanding and the Development of Intellectual Values," *International Journal of Educational Research* 43 (2005).

55. Barzilai and Chinn, "On the Goals of Epistemic Education."

56. Jason Baehr, "Educating for Intellectual Virtues: From Theory to Practice," *Journal of Philosophy of Education* 47, no. 2 (2013).

57. NASA, "What Is Climate Change?" https://climate.nasa.gov.

References

Baehr, Jason. "Educating for Intellectual Virtues: From Theory to Practice." *Journal of Philosophy of Education* 47, no. 2 (2013): 248–61.

Barzilai, Sarit, and Clark A. Chinn. "On the Goals of Epistemic Education: Promoting Apt Epistemic Performance." *Journal of the Learning Sciences* 27 (2018): 353–89.

BBC Trust. "Review of Impartiality and Accuracy of the BBC's Coverage of Science." https://www.bbc.co.uk/bbctrust/our_work/editorial_standards/impartiality/science_impartiality.html.

Beach, Josh M. *How Do You Know? The Epistemological Foundations of 21st Century Literacy*. New York: Routledge, 2018.

Boseley, Sarah. "How Disgraced Anti-Vaxxer Andrew Wakefield Was Embraced by Trump's America." *The Guardian*, July 18, 2018.

Boykoff, Maxwell T., and Jules M. Boykoff. "Balance as Bias: Global Warming and the US Prestige Press." *Global Environmental Change* 14, no. 2 (2004): 125–36.

Bramoullé, Yann, and Caroline Orset. "Manufacturing Doubt." *Journal of Environmental Economics and Management* 90 (2018): 119–33.

Bromme, Rainer, and Susan R. Goldman. "The Public's Bounded Understanding of Science." *Educational Psychologist* 49, no. 2 (2014): 59–69.

Burdick, Alan. "Looking for Life on a Flat Earth: What a Burgeoning Movement Says About Science, Solace, and How a Theory Becomes Truth." *The New Yorker*, May 30, 2018.

Centers for Disease Control and Prevention. "Opioid Overdose: Understanding the Epidemic." https://www.cdc.gov/drugoverdose/epidemic/index.html.

CNN. "19 States See Rise in Covid-19 Cases Amid Reopening and Protests." June 12, 2020.

Duncan, Ravit G., Clark A. Chinn, and Sarit Barzilai. "Grasp of Evidence: Problematizing and Expanding Next Generation Science Standards' Conceptualization of Evidence." *Journal of Research in Science Teaching* 55 (2018): 905–37.

Evans, John H., and Eszter Hargittai. "Why Would Anyone Distrust Anthony Fauci?" *Scientific American*, June 7, 2020. https://blogs.scientificamerican.com/observations/why-would-anyone-distrust-anthony-fauci/.

Felton, Mark K., and Deanna Kuhn. " 'How Do I Know?' the Epistemological Roots of Critical Thinking.' " *Journal of Museum Education* 32 (2007): 101–10.

Funk, Cary, Meg Hefferon, Brian Kennedy, and Courtney Johnson. "Trust and Mistrust in Americans' Views of Scientific Experts." Pew Research Center, August 2, 2019. https://www.pewresearch.org/science/2019/08/02/trust-and-mistrust-in-americans-views-of-scientific-experts/.

Grasswick, Heidi E. "Scientific and Lay Communities: Earning Epistemic Trust Through Knowledge Sharing." *Synthese* 177, no. 3 (2010): 387–409.

Hendriks, Friederike, Dorothe Kienhues, and Rainer Bromme. "Trust in Science and the Science of Trust." In *Trust and Communication in a Digitized World*, edited by Bernd Blöbaum, 143–59. Cham, Switzerland: Springer, 2016.

Hofer, Barbara K. "Personal Epistemology Research: Implications for Learning and Teaching." *Educational Psychology Review* 13 (2001): 353–83.

Hofer, Barbara K. "Beliefs About Knowledge and Knowing: Integrating Domain Specificity and Domain Generality: A Response to Muis, Bendixen, and Haerle (2006)." *Educational Psychology Review* 18, no. 1 (2006): 67–76.

Hofer, Barbara K. "Epistemic Cognition as a Psychological Construct: Advancements and Challenges." In *Handbook of Epistemic Cognition*, edited by Jeffrey Alan Greene, William A. Sandoval and Ivar Bråten, 19–38. New York: Routledge, 2016.

Hofer, Barbara K., C. F. Lam, and A. DeLisi. "Understanding Evolutionary Theory: The Role of Epistemological Development and Beliefs." In *Epistemology and Science Education: Understanding the Evolution vs. Intelligent Design Controversy*, edited by R. Taylor and M Ferrari, 95–110. New York: Routledge, 2011.

Hofer, Barbara K., and Paul R. Pintrich. "The Development of Epistemological Theories: Beliefs About Knowledge and Knowing and Their Relation to Learning." *Review of Educational Research* 67, no. 1 (1997): 88–140.

Kimmerle, Joachim, Danny Flemming, Insa Feinkohl, and Ulrike Cress. "How Laypeople Understand the Tentativeness of Medical Research News in the Media: An Experimental Study on the Perception of Information About Deep Brain Stimulation." *Science Communication* 37, no. 2 (2015): 173–89.

Koehler, Derek J. "Can Journalistic 'False Balance' Distort Public Perception of Consensus in Expert Opinion?" *Journal of Experimental Psychology: Applied* 22 (2016a): 24–38.

Koehler, Derek J. "Why People Are Confused About What Experts Really Think." *New York Times*, February 14, 2016b.

Kuhn, Deanna, Richard Cheney, and Michael Weinstock. "The Development of Epistemological Understanding." *Cognitive Development* 15, no. 3 (2000): 309–28.

Kuhn, Deanna, and Seung-Ho Park. "Epistemological Understanding and the Development of Intellectual Values." *International Journal of Educational Research* 43 (2005): 111–24.

Liu, Katherine A., and Natalie A. D. Mager. "Women's Involvement in Clinical Trials: Historical Perspective and Future Implications." *Pharmacy Practice* 14 (2016): 708.

Lombardi, Doug. "Thinking Scientifically in a Changing World." American Psychological Association, January 2019. https://www.apa.org/science/about/psa/2019/01/changing-world.aspx.

McIntyre, Lee. *Post-Truth*. Cambridge, MA: MIT Press, 2018.

McIntyre, Lee. *The Scientific Attitude: Defending Science from Denial, Fraud, and Pseudoscience.* Cambridge, MA: MIT Press, 2019.

NASA. "What Is Climate Change?" https://climate.nasa.gov.

National Research Council. *A Framework for K–12 Science Education: Practices, Crosscutting Concepts, and Core Ideas.* Washington, DC: National Academies Press, 2012.

NGSS Lead States. *Next Generation Science Standards: For States, by States.* Washington, DC: National Academies Press, 2013.

Nichols, Tom. *The Death of Expertise: The Campaign Against Established Knowledge and Why It Matters.* New York: Oxford University Press, 2017.

Oreskes, Naomi. *Why Trust Science?* Princeton, NJ: Princeton University Press, 2019.

Oreskes, Naomi, and Erik M. Conway. *Merchants of Doubt: How a Handful of Scientists Obscured the Truth on Issues from Tobacco Smoke to Global Warming.* New York: Bloomsbury Publishing, 2010.

Orr, Daniela, and Ayelet Baram-Tsabari. "Science and Politics in the Polio Vaccination Debate on Facebook: A Mixed-Methods Approach to Public Engagement in a Science-Based Dialogue." *Journal of Microbiology and Biology Education* 19 (2018): 1–8.

Rabin, R. C., and E. Gabler. "Two Huge Covid-19 Studies Are Retracted After Scientists Sound Alarms." *New York Times*, June 4, 2020.

"Retraction: Study on Chloroquine and Hydroxychloroquine in COVID-19 Patients." *The Lancet* May 22, 2020.

Shapiro, Jeremy P. "The Thinking Error at the Root of Science Denial." *The Conversation*, May 8, 2018. https://theconversation.com/the-thinking-error-at-the-root-of-science-denial-96099.

Sinatra, Gale M., and Barbara K. Hofer. "Public Understanding of Science: Policy and Educational Implications." *Policy Insights from the Behavioral and Brain Sciences* 3, no. 2 (2016): 245–53.

Stocking, S. Holly, and Lisa W. Holsteing. "Manufacturing Doubt: Journalists' Roles and the Construction of Ignorance in Scientific Controversy." *Public Understanding of Science* 18 (2009): 23–42.

Stoknes, Per Espen. *What We Think About When We Try Not to Think About Global Warming: Toward a New Psychology of Climate Action.* White River Junction, VT: Chelsea Green Publishing, 2015.

6

What Motivates People to Question Science?

Beverly's husband had to stay home from his job in the coal mines of western Appalachia with an extreme coughing fit again. She feared that, like her father, Tom was in the early stages of black lung, a disease caused by coal dust that is common in miners, although he says it's probably just the flu. Coal mining is a good-paying job, and Tom had made a decent living before he got laid off when the economy tanked. Now that he's back at work, at least when he's not home sick, Beverly is getting caught up on the household bills. She is well aware that coal mining is dangerous work; her grandfather had died in a coal mining accident when she was just 6 years old. Now all these scientists are saying that burning coal contributes to global warming, but her husband says that's just political talk.

Beverly saw her father suffer from black lung, and she sure hopes Tom is right about the flu. Maybe coal mining is dangerous, but is it really contributing to global warming? Humans couldn't possibly affect the climate, could they? Beverly would rather see Tom doing something other than mining because of his health, but what would that be? No other job in their town pays half as well as mining. All the mining families she knows are convinced that the answer is clean coal. Beverly has seen enough coal in her life to realize it's a lot of things, but clean doesn't seem to be one of them. She wonders why everyone who makes a living in mining is so sure about clean coal, and those environmentalists, who have never worked a job like mining, seem so hell bent on keeping Tom unemployed. What's in it for them? Everyone seems to have a dog in this fight.

In the 2016 presidential campaign, one candidate avowed, "I believe in science" as part of a political platform, and another denied climate science altogether as a hoax. The United States pulled out of the Paris Climate Accord in 2017 and stood alone as the only member of the G-20 not to support previously agreed upon standards for addressing this global problem. In the spring of 2020, the video *Plandemic* reached millions of viewers, spreading

Science Denial. Gale M. Sinatra and Barbara K. Hofer, Oxford University Press. © Oxford University Press 2021.
DOI: 10.1093/oso/9780190944681.003.0006

misinformation about the novel coronavirus that had been championed by anti-government conspiracy theorists.[1] From stem cell research to genetically modified organisms (GMOs) to protective mask-wearing during a pandemic, battle lines are drawn around scientific issues, often with evidence taking a back seat, being ignored, or being misrepresented. Positions on both sides of these debates often reflect less about familiarity with the scientific topic than they do about individuals' goals and desires, leading to what psychologists call a "motivated view" of science. Even when individuals attempt to make decisions justified with evidence, motivations can bias what information they attend to and how much credence they give it, as well as the strategies they use to assess that information.

Motivations are essentially the goals or desires that move individuals toward or away from activities.[2] The desire to look great at an upcoming high school reunion can motivate someone toward an activity (such as going to the gym) or away from an activity (such as avoiding a bakery shop). Motivations are often positive. You can be motivated to get into shape, go back to school, or work toward that promotion. They can also be negative, such as the motivation to do harm or commit a crime. Motives also influence who people associate with and how they think. Motivations can influence the reasoning process itself by tilting a reasoner toward a desired conclusion, as attention is directed toward information that supports a preferred view and away from thoughtful consideration of contradictory data.

Individuals can have a variety of motives that impact their reasoning about science. Motivations can be economic, such as the motivation to maintain a job in the fossil fuel industry. They can be social, such as the motivation to continue to bring your child to their favorite play group, even if the moms disagree with your decisions regarding vaccines. A politician might be motivated to get re-elected by being dismissive of climate change in a district dominated by climate skeptics, or a social media influencer might be motivated to get "likes" for climate activism simply to increase their following. People have a variety of motives and goals, and these may even be competing. For example, individuals generally have a goal to survive and thrive, but they do not always take actions that are in their self-interest.

Political affiliations can also serve to motivate reasoning about science. Many have noted that erosion of trust in science has tended to follow along political party lines in recent decades, with conservatives reducing their trust in science and liberals retaining theirs.[3] Although this pattern is strong,[4] several known motivational biases operate similarly across the political spectrum.[5] As an example, Kahan and his colleagues devised a clever experiment that illustrates how individuals can be biased in their evaluation of data toward a desired interpretation.[6] Participants evaluated complex (and fabricated) data about the efficacy of a new skin treatment based on variations in use and outcomes and were asked to interpret whether the cream was effective. Two versions of data were used with two different groups of participants. Some people saw a version of the data that showed the skin cream was effective, while others saw data showing that it was not. The researchers then relabeled the same data as depicting the rise and fall of crime rates in communities that either did or did not enact concealed carry gun laws. Again, they created two versions, one showing the new law was associated with lower crime rates, and the other showing it was not.

When conservatives and liberals reasoned about the efficacy of a skin cream treatment, only reasoning ability (how facile participants were at evaluating data) mattered, but political affiliation did not. However, when conservatives and liberals reasoned about the fictitious crime rate data, the impact of political affiliation was more influential than reasoning skills, especially so when the data were in conflict with politically held beliefs. Kahan and colleagues argued that when individuals read about a topic for which they have no particular desired conclusion (skin treatment effectiveness, for example), they did as well as their reasoning abilities allowed. But when they had to reason about the complexities of gun regulations and crime rates, their prior beliefs that guns either keep you safe or that guns are a serious public health threat influenced their interpretations of the data. Their conclusions tended to align with how they viewed the world *should be*, rather than how it actually is. Participants who believed guns made you safer were motivated to see the data as supporting this conclusion, while the opposite was true for those who believed guns to be a public safety threat. This is quintessential motivated reasoning, allowing factors that should be irrelevant (such as desire to see support for a particular view) to interfere with reasoning about evidence.

Personal motivations on any side of an issue have the potential to bias the interpretations of hard data. As concerns about the mounting economic crisis escalated during 2020, whether interpretations of COVID-19 public health data were seen as supporting either gradual or rapid reopening of the economy was explained by political affiliation as much as (if not more than) public health best practice.[7] Beverly may be skeptical of climate change, but she is right to wonder about the motivations of others. Some environmentalists who accept the scientific research on climate change may be motivated to be critical of the science behind nuclear energy, for example, although it is seen by others as a reasonable alternative to fossil fuels for reducing environmental impact.[8] As Beverly wondered, if you have a dog in the fight, maybe it influences how you view the data.

How Does Motivated Reasoning Exert Its Influence?

Everyone likes to think of themselves as behaving in an unbiased fashion most of the time. We all view ourselves as fair and impartial arbiters of facts akin to the blindfolded statue of Lady Justice evaluating competing claims without bias, emotions, or motivations. And yet, as discussed in Chapter 4, overwhelming psychological research suggests that such unbiased rationality is actually a fairly elusive quality in humans.[9] Much of the time people are on automatic pilot. In other words, individuals are *acting* without reflection more often than they are *thinking* carefully and deliberately.[10] The rest of the time, even as individuals are trying their best to think through issues, motivational goals may bias their thought processes and bias their reasoning. Psychologist Ziva Kunda, who coined the term "motivated reasoning" to describe this phenomenon,[11] explained that although individuals try to make well-thought-out decisions, use available evidence, and look at both sides of an issue, the process is often tainted by motivations that may be unknown to them. Individuals' motivations may direct them to attend more carefully to some information while ignoring other relevant facts. Or they may use different strategies to evaluate information they prefer to be correct (such as dismissing obvious flaws) while at the same time being hypercritical of flaws in information they prefer to be wrong.

Individuals are more rigorous in their evaluation of evidence and claims that are in conflict with their own positions than those that are in accord with their prior attitudes. A classic study by Lord and his colleagues had individuals who were either in favor or against capital punishment read articles describing social science research that showed either support or no support for capital punishment as a crime deterrent. Individuals tended to be more critical of studies whose outcomes they did not favor compared to studies supporting their prior point of view.[12] They cited potentially legitimate methodological concerns such as insufficient sample size or lack of adequate controls or sampling errors. These reasonable research study critiques were identified much less frequently for studies that supported their views. The researchers concluded that individuals' evaluation of research quality rested more on the alignment between the reasoners' beliefs and the study's conclusions than on the study's methodological rigor. Other research has also found the same biased evaluation of political arguments. For example, Charles Taber and Milton Lodge found that citizens evaluated arguments about affirmative action and gun control as stronger when they concurred with the position being taken and weaker when they disagreed.[13] They also found that individuals were far more likely to come up with counterarguments against those positions that were contrary to their own.

How could that happen? Kunda explains that individuals can have what she called a "directional goal"; that is, they can be motivated to arrive at a favored conclusion. This can happen to individuals who have a stake in the outcome, either personally, professionally, or financially. Beverly's friends and family may be skeptical of climate change (and coal's contribution) for a variety of reasons, including their personal and financial stake in the continuation of a livelihood in the fossil fuel industry. Similarly, a solar panel salesperson may be motivated to overestimate the benefits of solar energy if paid by commissions. An environmental activist may put more weight on nuclear disasters than their numbers warrant when compared to their potential to reduce global warming. Someone who owns a restaurant or barbershop may view public health data on stay-at-home policies during a pandemic through the lens of the impact on their business. Anyone with a favored outcome may be compromised in their reasoning.

Evidence has been mounting that everyone is susceptible to the siren's song of their own motivations. Researchers in the fields of decision-making, problem-solving, and reasoning have argued for decades that individuals were motivated toward what Kunda calls an "accuracy goal," in other words, to be correct. Researchers may have given too much credit to reasoners, assuming that after making the effort to think through an issue, they would be motivated to be correct, given the information at hand and the problem-solving strategies at their disposal. However, individuals' tendencies to reason toward a preferred conclusion are much stronger than previously assumed.

Motivation can influence not only the selection of information that individuals attended to when reasoning but also the strategies that they use to think and reason with the evidence. When individuals are motivated to be accurate in their conclusions, they put forth the effort to weigh both sides of the issue. They are more likely to slow down their thinking and resist jumping to premature conclusions. However, if individuals are motivated (either knowingly or unwittingly) to arrive at a desired conclusion, they put forth extra effort to recall knowledge they can use to refute information that is a threat to their position. At the same time, these individuals are prone to ignore evidence that contradicts their preferred conclusion.

Everyone can fall prey to motivated reasoning, in part because the effort it takes to be accuracy-driven is substantial. Kunda acknowledged there is a cost/accuracy trade-off. Imagine that Martin is concerned about climate change, but he is not sure whether it is primarily driven by natural causes or a result of human activity. If Martin is accuracy-driven, he may try to find information that argues for both causes and evaluate that information to the best of his abilities. If he discusses what he finds online with his friend Ellie, who has heard that climate change is a hoax, she may not be accuracy-driven. If instead Ellie is motivated by her directional goal to come to her preferred conclusion, she might remind Martin about "Snowmageddon," the megastorm in 2010 hit the US East Coast, as evidence against global warming. But she might not mention (or even recall) that sea level rise worsened the impact of Hurricane Sandy in 2012, a result many scientists think was exacerbated by climate change. Martin might find Ellie so convincing that he questions whether he found the best information online. Obtaining a fair and accurate set of information with which to judge the plausibility of scientific claims can be difficult for non-scientists.[14] Moreover, the interpretation of that information can be strongly influenced by those in our media or social circles.

Researchers explain that most individuals do not usually just make up facts or draw conclusions that are untethered to reality (although this trend might be changing as such behaviors have become more frequent in political discourse). Rather, more often, they create a justification for their conclusions that is persuasive to the casual observer. For example, Maya Goldenburg, a philosopher interested in the public understanding of science, explains that one of the many reasons why new parents are compelled to believe that vaccines cause autism is that they hear stories from other parents about how their child's first symptoms appeared right after they completed the course of vaccinations.[15] This proximity in time appears to promote the misconception that there is a causal link between vaccinations and autism. It is an unfortunate coincidence that when a child with autism begins to exhibit language delays (around the age of 2) is precisely the same time the initial childhood vaccination sequence wraps up.[16] David Armor called this maintaining "an illusion of objectivity."[17] That is, if you examine the time course, you see the proximity of these two events, and it feels as though you have done a careful and objective analysis. This illusion of unbiased examination of the facts cannot be maintained indefinitely. At some point, the weight of the evidence does become overwhelming, if a person is open to it.

Take, for example, the case of geoscientists who succumbed to what may have been motivated reasoning about plate tectonics. As Naomi Oreskes and Homer LeGrand recount in *Plate Tectonics: An Insider's History of the Modern Theory of the Earth*,[18] in the early 1900s, geoscientists strongly resisted Alfred Wegener's theory that the continents shifted around the planet for millennia. The new science of "continental drift," or what is now called "plate tectonics," took over 40 years to be accepted as mainstream science, largely due to resistance among the geoscientists who held the traditional "stasis" view. Scientists may have evaluated the data on continental drift with a desire to confirm their preferred view, which was that the continents stayed put. This was the accepted view at the time, and for many reasons (careers, status in the academy) some geoscientists were less than objective about the data supporting the continents shifting beneath their feet.

However, despite decades of pushback from the stasis-view adherents, plate tectonics was eventually fully accepted. In fact, it is a foundational textbook concept within geoscience today. Evidence can and often does win over motivated reasoning in the long run, but major scientific revolutions of thought, like continents, can move very slowly until there is a seismic shift.

Motivated Reasoning When Learning About Science Online

Imagine Sarah is a teen who is concerned about the safety of vaping. She may do her due diligence by going online to research vaping safety. There, she will come across a large number of articles with conflicting information. If she is accuracy-driven, she will try to fairly evaluate both sides of the evidence. Now, imagine Sarah has heard from her friends that vaping is much safer than smoking. The teen might come across a number of articles that make this claim, while still many other articles refute it. If she is motivated to come to a particular conclusion, that vaping is safer than smoking, she might weigh articles suggesting that vaping is safe more heavily than those suggesting that vaping has serious health risks. Since motivated reasoning is not a deliberate act, she would likely be unaware of this tendency to favor one perspective over the other. The challenge is knowing whom to trust when seeking scientific information and how to evaluate that information critically and not with an eye toward supporting a favored conclusion.

Motivated reasoning is faulty reasoning. Yet members of the public rarely have the opportunity to evaluate firsthand evidence or have the means to directly confirm scientific data easily. Therefore, it really matters that individuals know how to find and evaluate scientific sources. There has been a dire need to evaluate scientific information found online since the COVID-19 outbreak. Information, misinformation, and disinformation can spread faster than the pandemic across online news and social media platforms. Gale and her colleague Doug Lombardi described a number of steps individuals should take to critically evaluate scientific information online, such as fact-checking information about how the virus spreads before sharing it on social media and judging whether the information seems plausible.[19] An individual who is accuracy-driven in their goals to evaluate scientific information has to make a concerted effort to confirm or disconfirm information. This takes time and effort to obtain, read, and evaluate information to ascertain what is reliable and valid information and what is not.

Social Identity

Clearly, thinking and reasoning can be biased by motivations. However, individuals do not just think and reason alone. Human beings are social creatures. Everyone feels as though they belong in some groups but do not belong in others. People identify with others when they share something in common, such as interests, hobbies, political affiliation, gender, religion, ethnicity, age, or socioeconomic status; these individuals are members of their in-groups. Not all groups you belong to are equally important or meaningful. However, in-groups that hold psychological meaning for *social identity* can be influential in decision-making.[20] A social identity is how individuals see themselves as members of a particular social group (e.g., a woman, a teacher, a mother, an Italian American). Individuals tend to conform to the attitudes, norms, and behaviors of those in their in-group, and often these social identities are further defined in contrast to the attitudes, norms, and behaviors of an out-group. For example, in terms of political in-group and out-group views on science, Democrats tend to be accepting of stem cell research, evolution, and climate change; but they may be skeptical of nuclear energy and genetically modified organisms. Republicans tend to be relatively more accepting of nuclear energy and genetically modified organisms but skeptical of stem cell research, evolution, and climate change. Individuals trying to make a decision about whether to support a particular candidate due to their climate change policy might weigh the opinion of those in their social group more heavily than those of climate scientists or policy makers.

The development of a social identity and group affiliation has two key features. First, one key aspect is the psychological sense of belonging, seeing oneself as part of a community. The second key aspect is social acceptance. It is important that the community recognize the individual as a group member who fits in.[21] If this all sounds a little like high school, it should. Adolescents are busy developmentally sorting out their own personal psychological identities, and thus they are keenly sensitive to group belonging and social acceptance.[22] But it is not just teenagers who define themselves by their associations—group affiliation is a trait that is deeply rooted in our evolutionary history and a defining feature of what makes us human. In our ancient past, being part of a group that "had your back" was key to survival. Knowing who was a member of a group that could not be trusted was also a matter of life or death. Those in-groups and out-groups traditionally formed by virtue of when and where one was born and lived and probably remained relatively stable throughout individuals' lives.

Today, fueled by demographic and technological shifts in US culture, in-groups and out-groups are formed quickly and easily, often by choice more than accident of birth. Author Bill Bishop, in *The Big Sort: Why the Clustering of Like-Minded America Is Tearing Us Apart*,[23] described how individuals in the United States have been sorting themselves into more and more homogeneous communities by deliberately relocating to specific places. In the deep past of human history, and right through much of modern times, individuals tended to remain close to the community where they were born. Bishop traced how, for the past several decades, Americans who have the means and the opportunity have been systematically choosing to relocate to areas where there are like-minded social groups. Consider how the community of Austin, Texas, is much more liberal than the surrounding state.[24] Since Bishop's analysis, this trend of surrounding ourselves with like-minded social circles has only been magnified by the explosion of social media, where individuals have been electronically sorting themselves into in-groups and out-groups more efficiently and effectively than ever before, without having to move to another city or town.

Social identity is reflected by the attitudes (e.g., urban residents tend to be more liberal than rural residents), traits and stereotypes (e.g., scientists are geeky), and behaviors (e.g., professors give lectures) that represent the group. The *prototype* is the individual who is the most typical representation of their social group.[25] Gale and her colleagues have described how some women struggle to develop a view of themselves as scientists. This is due in part to a deeply rooted societal view of a typical scientist as "White, male, socially awkward, and singularly obsessed with their chosen [science] field."[26] For example, the popular show *The Big Bang Theory* capitalizes on this prototype of a physicist as most of the cast is male and socially awkward.

If joining a social group is easier than ever before due to social and electronic mobility, the same is not necessarily true of leaving a social group. Social groups exert powerful influences over their members to keep them in line. In the extreme, cults are defined by their willingness to take extraordinary measures to keep members in the fold. Even a mothers' support group may put up some resistance when they hear that you aren't planning to breastfeed your infant, regardless of your own reasoning. Groups that matter most are those that individuals strongly associate with as part of their identity, such as political or religious groups. These groups are difficult to leave in part because members encourage other members to both stay in the fold and share similar points of view. Social groups were once critical to human survival and still play a powerful social and psychological role in individuals' lives, often for the betterment of their members. Social isolation leads to loneliness, depression, and despair. The rising suicide rate in the United States among teens has been attributed in part to increased perceptions of social isolation or ostracization from online social groups.[27]

Not surprisingly, research suggests that messages coming from a member of an in-group are much more persuasive than the same messages coming from outsiders.[28] Taking the received wisdom of the group on a scientific issue is quicker and easier than doing the research yourself. If a middle-aged woman wants to know whether it is a good idea for her to go on hormone replacement therapy, she might ask her friends what they think. Or individuals might crowdsource opinions on political issues. In California, where there are often dozens of ballot measures, it is not uncommon to see on social media, "Hey, how is everybody voting on Proposition 12?" As busy people, most individuals are inclined to accept their group's view on a topic, rather than conduct the laborious, time-consuming research needed to gather relevant information. However, when individuals crowd-source their views on science by polling friends, rather than evaluating science claims on their own merit, they are not as likely to make scientifically sound decisions.

Individuals who are not motivated to do the research or do not want to invest the time to do a deep dive into the complex nuances of a scientific debate might take the low-effort route of accepting the thinking of their in-group on science topics. One might think, "I don't know much about GMOs, but all my friends are avoiding GMOs, so I guess GMOs are bad." The power of the anti-gluten, anti-GMO, anti-vaccine in-groups in Southern California led the talk show host Jimmy Kimmel to famously quip, "Here in LA, there are schools in which 20% of the students aren't vaccinated because parents are more scared of gluten than they are of smallpox."[29]

Forming our views on science through the lens of social groups is problematic for several reasons. First, accepting or denying scientific issues based on what one's social group thinks can truncate critical thinking about those issues. The decision to wear a mask during a pandemic should be based on whether masks are effective in mitigating the spread of a virus, not on what your in-group thinks about mask-wearing. Second, if the plausibility of a scientific explanation depends on particular scientists' group affiliations rather than the evidence, then trust in science as an endeavor that seeks reliable but not necessarily comfortable answers would be undermined. Finally, the notion that junk science is whatever is inconsistent with one's social group's preferences reinforces a negative perception. It suggests that science, as an enterprise, is untrustworthy.

Returning to Beverly and her concerns about the contributions of coal to climate change, like all of us, Beverly is not thinking alone in an isolated reasoning chamber. Beverly is listening to her husband Tom's view and that of her other family members. She is discussing issues with her friends, who are shaping her point of view. She is reading newspapers, doing research online, listening to cable news, and engaging with friends on social media. If she's like most Americans, those information sources tend to be aligned with the values of her social group, reinforcing what she already believes and making it difficult to find disconfirming evidence.

The power of an in-group is at its peak when the issue is deeply polarized. For example, climate science acceptors and deniers form opposing camps, with members on each side vigorously endorsing pro– or anti–climate change messages. This is why you don't often hear politicians saying, "climate change is a complex and nuanced socio-scientific issue which requires careful consideration of all sides of the issue." Instead, you hear messages such as, "climate change is a hoax" or "climate change is an existential threat to civilization." Such statements serve to maximally differentiate the groups and signal the politician's view, but they do little to help individuals evaluate the evidence that would lead to clarity of understanding.

One downside of a strong social group with a clear point of view is that it is difficult to buck the group's thinking, even when it is not supported by evidence. In extreme cases, group members can be rather assertive in keeping members aligned with their shared values. It is considered a newsworthy event if a Republican leader comes out in support of climate science or a member of the National Rifle Association speaks out about gun violence. Those who break philosophically with the group risk being ostracized. For example, a young mother might be reticent to tell her anti-vaccine friends that she has decided to stick with her doctor's recommended vaccination schedule for her infant. She may fear questioning of her judgment as a parent or worry about being ostracized by the group. If group members are reluctant to share out-group views that contradict the beliefs of in-group members, the chance to discuss pros and cons on any topic is limited. If Beverly becomes interested in alternative energy, even if her view is a deeply informed one, she may find herself on the outside of her social group if she shares the desire to put solar panels on her roof.

Both of us (Gale and Barbara) have done research in the area of evolution acceptance and find that accepting the scientific evidence puts some individuals into a state of conflict with their social group (their family members or church group), and for others it may also lead them to question their own identity. As described in Chapter 5, a first-year student in Barbara's study questioned how his family of creationists would react to his new views on evolution, calling his identity into question, an issue Gale has also heard students raise in her interview studies. Identity statements are powerful reminders that beliefs about scientific topics that are perceived to be in conflict with deeply held ideas about the self can be strongly resistant to change.

When they do change, individuals may feel that their social identities are threatened.

In such circumstances, change is much more likely to occur if individuals hear from an in-group member. That is why Elizabeth Barnes and Sarah Brownell from Arizona State University have been including video messages from Christian biologists to share with their undergraduate biology students.[30] Having someone from your in-group who is also an expert in the field allows science students to see that their identity (as members of a particular faith) can remain intact, even as they come to accept the scientific perspective.

In Gale's research with her colleagues on attitudes about scientific views on climate change, she found a similar pattern: members of an in-group were more persuasive in communicating about the science than members of an out-group,[31] a finding consistent with other research which shows that in-group members consult the members of their group who hold similar worldviews when it comes to deciding their attitudes toward a scientific topic.[32] However, in Gale's study, they found a pattern that was the opposite of the typical finding that in-group members are more persuasive. When it came to knowledge-based judgments (verifiably right or wrong), not opinions, out-group members did hold some sway. So Beverly might be influenced by the attitudes of her in-group toward alternative energy, but for an explanation of how solar panels absorb and transfer energy back to the grid, an electrical engineer might be judged as more knowledge-based and less biased than an in-group member known to be against solar.

In sum, individuals see their membership in a group as part of who they are, so much so that their identity and their social relationships are often defined by their positions. Consider those people who proudly claim "I'm an environmentalist" or "I'm anti-GMOs" as making a statement of who they are and where they belong. These identities can be pro– or anti–scientific attitudes. Identity-based views of science (be they on the left or right of the political spectrum) can lead to biased reasoning just as easily as the motivations discussed at the beginning of this chapter.

What Can We Do?

When individuals are motivated toward a particular outcome, reasoning about the topic is less than optimal. We reviewed research in Chapter 4 that shows that suppressing our motives and biases to think critically about a complex issue requires mental effort, especially when the issues pose strong conflicts with our own ideas about how the world works. Similarly, reflecting on our motivations to believe or reject scientific information does take concerted attention, but it is the first step toward accuracy-driven decision-making.

What Can Individuals Do?

Take the time and effort to reflect on your own preferred outcome before jumping to the internet or reasoning about a position on a scientific topic. Stop, step back, and ask yourself, *What do I want to believe?* Then, armed with that awareness, ask, *What does the best evidence suggest?* And finally ask, *Are those two positions aligned?* If they are not aligned, give yourself the chance to consider the position as fairly as you can before you decide. You can also consider what sources to examine before you seek information. If you have concerns about the safety of nuclear energy, watching the TV miniseries *Chernobyl* may not be your best source of scientifically accurate information and will likely only serve to reinforce your existing concerns.

Gale and her colleague Doug Lombardi have written about the challenges of identifying the source of scientific information when reading online in the "fake news" era.[33] Purveyors of misleading or false information about science have become increasingly sophisticated in how they present information on professionally designed websites that quote "experts" who look and sound like scientists but may or may not be reputable.

All of us should be aware of how our social identity shapes our thinking and willing to consider points of view that our social group members do not endorse. Gale's health-conscious neighbor made her an algae smoothie, promoting the health benefits of daily consumption, but it actually made her sick. A review of the side effects posted on the Centers for Disease Control and Prevention showed not only did algae give some people headaches and nausea, but that people with autoimmune disorders like Gale's should avoid consuming algae. To make sound decisions, it is better to evaluate the evidence than just run with the herd. Individuals need to get out of their information bubbles to hear new ideas and be open to alternative points of view and willing to engage with them fairly.

In one of Gale's own research studies, she and her colleague asked college students to argue in favor of a view they did not hold but which was scientifically accurate.[34] Students were asked "You picked A, but can you imagine why someone would pick B"? (Response A was incorrect, and B was the correct scientific answer.) After they gave an explanation of why someone might choose B (the correct response), they were asked, "OK, now, do you want to stick with your original response?" Students who were asked to argue in favor of the correct answer they had not originally selected were more likely to switch to endorsing that correct view. Similarly, a famous climate skeptic was asked to evaluate the evidence about climate change and when he did, he accepted that evidence.[35]

What Can Science Communicators Do?

For those who communicate about science, such as scientists writing about their work, science educators, and science writers, there are more and less successful ways to frame messages for a skeptical audience. Science communicators need to be keenly aware of their audiences' views and how they and their audience members identify. They need to consider what motivations individuals may have to accept or reject the information they are trying to share. Membership status of the science communicator is important because in-group communicators are often seen as more credible. If you are communicating to an out-group, consider drawing on an in-group ally to help frame or deliver your message. A doctor who is trying to communicate about vaccine safety to expectant mothers would be well served to include patients who recently vaccinated their child in those conversations to share their concerns, knowledge, and experiences.

Social psychologist Viviane Seyranian has shown in her work that inclusive messages, those that use "we" rather than "I" language, can appeal to identity in a positive way that mobilizes action.[36] Seyranian (together with Gale and a colleague) has shown that evoking individuals' identification with their community, for example saying, "we Southern Californians, we conserve water," can more effectively reduce water consumption than the typical comparative information water companies often provide consumers about their usage.[37] Famously, after 9/11, George Bush stood on a pile of rubble at ground zero and said that those who were responsible for this "would hear from all of *us* soon." The power of that message was the framing of "us" to mean all Americans.[38] For a brief moment, the most important in-group was Americans. Democratic candidates in the 2019 televised Climate Town Hall each said "I" accept or "I" would listen to the science of climate change. It would have actually been more powerful to say, "As Democrats, we accept the science on climate change."

Conclusions

Individuals' goals and motivations can influence reasoning about science. These motivations can be economic, social, personal, or political; and they are likely to affect what individuals attend to, how they process information, and the strategies they use to evaluate it. In addition, social identity can strongly shape views about science. Social identities are multifaceted, influential aspects of the self and an important component to acknowledge in understanding science doubt and denial.

Notes

1. Sheera Frenkel, Ben Decker, and Davey Alba, "How the 'Plandemic' Movie and Its Falsehoods Spread Widely Online," *New York Times*, May 21, 2020, https://www.nytimes.com/2020/05/20/technology/plandemic-movie-youtube-facebook-coronavirus.html.
2. P. Karen Murphy and Patricia A. Alexander, "A Motivated Exploration of Motivation Terminology," *Contemporary Educational Psychology* 25, no. 1 (2000).
3. Stephan Lewandowsky and Klaus Oberauer, "Motivated Rejection of Science," *Current Directions in Psychological Science* 25, no. 4 (2016).
4. Lauran Neergaard and Hannah Fingerhut, "AP-NORC Poll: Half of Americans Would Get a COVID-19 Vaccine," Associated Press, May 27, 2020, https://apnews.com/dacdc8bc428dd4df6511bfa259cfec44.
5. Dan M. Kahan, "Ideology, Motivated Reasoning, and Cognitive Reflection: An Experimental Study," *Judgment and Decision Making* 8 (2012).
6. Dan M. Kahan et al., "Motivated Numeracy and Enlightened Self-Government," *Behavioural Public Policy* 1, no. 1 (2017).
7. Jack Brewster, "87% of the States That Have Reopened Voted for Trump in 2016," *Forbes*, May 5, 2020, https://www.forbes.com/sites/jackbrewster/2020/05/05/87-of-the-states-that-have-reopened-voted-for-trump-in-2016/#290b25cd2426.
8. Brad Plumer, Henry Fountain, and Livia Albeck-Ripka, "Environmentalists and Nuclear Power? It's Complicated," *New York Times*, April 18, 2018, https://www.nytimes.com/2018/04/18/climate/climate-fwd-green-nuclear.html.

9. Keith E. Stanovich, *Who Is Rational? Studies of Individual Differences in Reasoning* (Mahwah, NJ: Lawrence Erlbaum Associates, 1999).

10. John A. Bargh and Tanya L. Chartrand, "The Unbearable Automaticity of Being," *American Psychologist* 54, no. 7 (1999).

11. Ziva Kunda, "The Case for Motivated Reasoning," *Psychological Bulletin* 108, no. 3 (1990).

12. Charles G. Lord, Lee Ross, and Mark R. Lepper, "Biased Assimilation and Attitude Polarization: The Effects of Prior Theories on Subsequently Considered Evidence," *Journal of Personality and Social Psychology* 37 (1979).

13. Charles S. Taber and Milton Lodge, "Motivated Skepticism in the Evaluation of Political Beliefs," *American Journal of Political Science* 50, no. 3 (2006).

14. Gale M. Sinatra and Doug Lombardi, "Evaluating Sources of Scientific Evidence and Claims in the Post-Truth Era May Require Reappraising Plausibility Judgments," *Educational Psychologist* 55, no. 3 (2020).

15. Centers for Disease Control and Prevention, "Preparing for Questions Parents May Ask About Vaccines," April 11, 2018, https://www.cdc.gov/vaccines/hcp/conversations/preparing-for-parent-vaccine-questions.html.

16. Maya J. Goldenberg, "Public Misunderstanding of Science? Reframing the Problem of Vaccine Hesitancy," *Perspectives on Science* 24, no. 5 (2016).

17. D. A. Armor, "The Illusion of Objectivity: A Bias in the Perception of Freedom from Bias," *Dissertation Abstracts International: Section B: The Sciences and Engineering* 59, no. 9-B (1999).

18. Naomi Oreskes and Homer E. LeGrand, *Plate Tectonics: An Insider's History of the Modern Theory of the Earth* (Boulder, CO: Westview Press, 2001).

19. Sinatra, G. M., and Lombardi, D. "Evaluating Sources of Scientific Evidence and Claims in the Post-Truth Era May Require Reappraising Plausibility Judgments," *Educational Psychologist* 55(3) (2020): 120–131.

20. Michael A. Hogg, "Social Identity Theory," in *Contemporary Social Psychological Theories*, ed. Peter J. Burke (Stanford, CA: Stanford University Press, 2006).

21. H. Tajfel and John C. Turner, "Dimensions of Majority and Minority Groups," in *Psychology of Intergroup Relations*, ed. S. Worchel and W. G. Austin (Chicago: Nelson-Hall, 1986).

22. Erik H. Erikson, *Identity: Youth and Crisis* (New York: WW Norton, 1968).

23. Bill Bishop, *The Big Sort: Why the Clustering of Like-Minded America Is Tearing Us Apart* (New York: Houghton Mifflin Harcourt, 2009).

24. Ross Ramsey, "Analysis: The Blue Dots in Texas' Red Political Sea," *Texas Tribune*, November 11, 2016, https://www.texastribune.org/2016/11/11/analysis-blue-dots-texas-red-political-sea/.

25. Dominic Abrams and Michael A. Hogg, "Collective Identity: Group Membership and Self-Conception," in *Blackwell Handbook of Social Psychology: Group Processes*, ed. Michael A. Hogg and R. Scott Tindale (Oxford and Malden, MA: Blackwell, 2001).

26. Ann Y. Kim, Gale M. Sinatra, and Viviane Seyranian, "Developing a Stem Identity Among Young Women: A Social Identity Perspective," *Review of Educational Research* 88, no. 4 (2018): 593.

27. Amy L. Burnett, "Internet Related to Suicide," *International Journal of Child Health and Human Development* 10, no. 4 (2017).

28. Diane M. Mackie, Leila T. Worth, and Arlene G. Asuncion, "Processing of Persuasive in-Group Messages," *Journal of Personality and Social Psychology* 58, no. 5 (1990).

29. Christina Nuñez, "Jimmy Kimmel: Parents Are More Afraid of Gluten Than Smallpox," Global Citizen, February 27, 2015, https://www.globalcitizen.org/en/content/jimmy-kimmel-parents-are-more-afraid-of-gluten-tha/.

30. M. Elizabeth Barnes and Sara E. Brownell, "Experiences and Practices of Evolution Instructors at Christian Universities That Can Inform Culturally Competent Evolution Education," *Science Education* 102, no. 1 (2018).

31. Doug Lombardi, Viviane Seyranian, and Gale M. Sinatra, "Source Effects and Plausibility Judgments When Reading About Climate Change," *Discourse Processes* 51, no. 1–2 (2014).

32. W. D. Crano, *The Rules of Influence: Winning When You're in the Minority* (New York: St. Martin's Press, 2012).

33. Sinatra and Lombardi, "Evaluating Sources of Scientific Evidence."

34. E. Michael Nussbaum and Gale M. Sinatra, "Argument and Conceptual Engagement," *Contemporary Educational Psychology* 28, no. 3 (2003).

35. KQED News, "UC Berkeley, Lawrence Lab Climate-Change Skeptic Now Says Global Warming Is Real," October 31, 2011, https://www.kqed.org/news/45249/uc-berkeley-lawrence-berkeley-lab-climate-change-skeptic-now-says-global-warming-is-real.

36. Viviane Seyranian, "Public Interest Communication: A Social Psychological Perspective," *Journal of Public Interest Communication* 1 (2017).

37. Seyranian, V., Sinatra, G. M., and Polikoff, M. "Comparing Communication Strategies for Reducing Residential Water Consumption," *Journal of Environmental Psychology* 41 (2015): 81–90.

38. Viviane Seyranian, "Social Identity Framing Communication Strategies for Mobilizing Social Change," *Leadership Quarterly* 25, no. 3 (2014).

References

Abrams, Dominic, and Michael A. Hogg. "Collective Identity: Group Membership and Self-Conception." In *Blackwell Handbook of Social Psychology: Group Processes*, edited by Michael A. Hogg and R. Scott Tindale, 425–60. Oxford and Malden, MA: Blackwell, 2001.

Armor, D. A. "The Illusion of Objectivity: A Bias in the Perception of Freedom from Bias." *Dissertation Abstracts International: Section B: The Sciences and Engineering* 59, no. 9-B (1999): 5163.

Bargh, John A., and Tanya L. Chartrand. "The Unbearable Automaticity of Being." *American Psychologist* 54, no. 7 (1999): 462–79.

Barnes, M. Elizabeth, and Sara E. Brownell. "Experiences and Practices of Evolution Instructors at Christian Universities That Can Inform Culturally Competent Evolution Education." *Science Education* 102, no. 1 (2018): 36–59.

Bishop, Bill. *The Big Sort: Why the Clustering of Like-Minded America Is Tearing Us Apart.* New York: Houghton Mifflin Harcourt, 2009.

Brewster, Jack. "87% of the States That Have Reopened Voted for Trump in 2016." *Forbes*, May 5, 2020. https://www.forbes.com/sites/jackbrewster/2020/05/05/87-of-the-states-that-have-reopened-voted-for-trump-in-2016/#290b25cd2426.

Burnett, Amy L. "Internet Related to Suicide." *International Journal of Child Health and Human Development* 10, no. 4 (2017): 335–38.

Centers for Disease Control and Prevention. "Preparing for Questions Parents May Ask About Vaccines." April 11, 2018. https://www.cdc.gov/vaccines/hcp/conversations/preparing-for-parent-vaccine-questions.html.

Crano, W. D. *The Rules of Influence: Winning When You're in the Minority.* New York: St. Martin's Press, 2012.

Erikson, Erik H. *Identity: Youth and Crisis.* New York: WW Norton, 1968.

Frenkel, Sheera, Ben Decker, and Davey Alba. "How the 'Plandemic' Movie and Its Falsehoods Spread Widely Online." *New York Times*, May 21, 2020. https://www.nytimes.com/2020/05/20/technology/plandemic-movie-youtube-facebook-coronavirus.html.

Goldenberg, Maya J. "Public Misunderstanding of Science? Reframing the Problem of Vaccine Hesitancy." *Perspectives on Science* 24, no. 5 (2016): 552–81.

Hogg, Michael A. "Social Identity Theory." In *Contemporary Social Psychological Theories*, edited by Peter J. Burke, 111–36. Stanford, CA: Stanford University Press, 2006.

Kahan, Dan M. "Ideology, Motivated Reasoning, and Cognitive Reflection: An Experimental Study." *Judgment and Decision Making* 8 (2012): 407–24.

Kahan, Dan M., Ellen Peters, Erica Cantrell Dawson, and Paul Slovic. "Motivated Numeracy and Enlightened Self-Government." *Behavioural Public Policy* 1, no. 1 (2017): 54–86.

Kim, Ann Y., Gale M. Sinatra, and Viviane Seyranian. "Developing a Stem Identity Among Young Women: A Social Identity Perspective." *Review of Educational Research* 88, no. 4 (2018): 589–625.

KQED News. "UC Berkeley, Lawrence Lab Climate-Change Skeptic Now Says Global Warming Is Real." October 31, 2011. https://www.kqed.org/news/45249/uc-berkeley-lawrence-lab-climate-change-skeptic-now-says-global-warming-is-real.

Kunda, Ziva. "The Case for Motivated Reasoning." *Psychological Bulletin* 108, no. 3 (1990): 480–98.

Lewandowsky, Stephan, and Klaus Oberauer. "Motivated Rejection of Science." *Current Directions in Psychological Science* 25, no. 4 (2016): 217–22.

Lombardi, Doug, Viviane Seyranian, and Gale M. Sinatra. "Source Effects and Plausibility Judgments When Reading About Climate Change." *Discourse Processes* 51, no. 1–2 (2014): 75–92.

Lord, Charles G., Lee Ross, and Mark R. Lepper. "Biased Assimilation and Attitude Polarization: The Effects of Prior Theories on Subsequently Considered Evidence." *Journal of Personality and Social Psychology* 37 (1979): 2098–109.

Mackie, Diane M., Leila T. Worth, and Arlene G. Asuncion. "Processing of Persuasive In-Group Messages." *Journal of Personality and Social Psychology* 58, no. 5 (1990): 812.

Murphy, P. Karen, and Patricia A. Alexander. "A Motivated Exploration of Motivation Terminology." *Contemporary Educational Psychology* 25, no. 1 (2000): 3–53.

Neergaard, Lauran, and Hannah Fingerhut. "AP-NORC Poll: Half of Americans Would Get a COVID-19 Vaccine." Associated Press, May 27, 2020. https://apnews.com/dacdc8bc428dd4df6511bfa259cfec44.

Nuñez, Christina. "Jimmy Kimmel: Parents Are More Afraid of Gluten Than Smallpox." *Global Citizen*, February 27, 2015. https://www.globalcitizen.org/en/content/jimmy-kimmel-parents-are-more-afraid-of-gluten-tha/.

Nussbaum, E. Michael, and Gale M. Sinatra. "Argument and Conceptual Engagement." *Contemporary Educational Psychology* 28, no. 3 (2003): 384–95.

Oreskes, Naomi, and Homer E. LeGrand. *Plate Tectonics: An Insider's History of the Modern Theory of the Earth*. Boulder, CO: Westview Press, 2001.

Plumer, Brad, Henry Fountain, and Livia Albeck-Ripka. "Environmentalists and Nuclear Power? It's Complicated." *New York Times*, April 18, 2018. https://www.nytimes.com/2018/04/18/climate/climate-fwd-green-nuclear.html.

Ramsey, Ross. "Analysis: The Blue Dots in Texas' Red Political Sea." *Texas Tribune*, November 11, 2016. https://www.texastribune.org/2016/11/11/analysis-blue-dots-texas-red-political-sea/.

Seyranian, Viviane. "Social Identity Framing Communication Strategies for Mobilizing Social Change." *Leadership Quarterly* 25, no. 3 (2014): 468–86.

Seyranian, Viviane. "Public Interest Communication: A Social Psychological Perspective." *Journal of Public Interest Communication* 1 (2017): 57–77.

Seyranian, V., Sinatra, G. M., and Polikoff, M. "Comparing Communication Strategies for Reducing Residential Water Consumption," *Journal of Environmental Psychology* 41 (2015): 81–90.

Sinatra, Gale M., and Doug Lombardi. "Evaluating Sources of Scientific Evidence and Claims in the Post-Truth Era May Require Reappraising Plausibility Judgments." *Educational Psychologist* 55, no. 3 (2020): 120–31.

Stanovich, Keith E. *Who Is Rational? Studies of Individual Differences in Reasoning*. Mahwah, NJ: Lawrence Erlbaum Associates, 1999.

Taber, Charles S., and Milton Lodge. "Motivated Skepticism in the Evaluation of Political Beliefs." *American Journal of Political Science* 50, no. 3 (2006): 755–69.

Tajfel, H., and John C. Turner. "Dimensions of Majority and Minority Groups." In *Psychology of Intergroup Relations*, edited by S. Worchel and W. G. Austin, 7–24. Chicago: Nelson-Hall, 1986.

7

How Do Emotions and Attitudes Influence Science Understanding?

Kenisha was 10 years old when she learned that Pluto was no longer considered a planet. Now she's in graduate school studying astronomy. Kenisha remembers her 5th-grade teacher explaining why Pluto, her favorite planet, wasn't a planet anymore. She couldn't understand how scientists could change the definition of a planet. She remembers feeling angry, confused, and sad. She had a long conversation with her parents over dinner about the reasons why scientists changed the categorization. She felt her whole world shifting because she thought scientific facts were unchanging. She worried what else she "knew" could just be changed tomorrow. She laughs now at the passions of her childhood self but also recognizes that those same passions drove her to study science. Kenisha's surprise at the shifting nature of knowledge in science led her to study it more in high school and ultimately become a science major in college. Now her emotions are focusing her energies on examining the evidence for water on Mars. She laughs at her childhood reaction to Pluto, but she gets it. "I see that same look in undergraduates taking their first astronomy class when I'm their teaching assistant. They get confused, even angry, when they hear things in class that conflict with what they thought they knew to be true. They get anxious when they hear the universe may be expanding, and they can get excited when we discuss whether there are other planets that have the conditions to support life. I think back on how Pluto's reclassification hit me so hard as a kid that it changed the course of my life as I ended up pursuing a career in astronomy. It makes me appreciate how they feel when they are learning about astronomy for the first time."

Kenisha's response to learning about Pluto's demotion in elementary school illustrates the variety of emotions experienced as people learn about complex scientific topics. Science is understood through the warm lens of emotions. Kenisha is like many others who experience a wide range of emotions as they think about scientific information learned about in the classroom or read

Science Denial. Gale M. Sinatra and Barbara K. Hofer, Oxford University Press. © Oxford University Press 2021.
DOI: 10.1093/oso/9780190944681.003.0007

about online. What does Kenisha's experience tell us about the role emotions play in the public understanding of science? How do emotions affect individuals' attitudes toward science, perhaps leading them to embrace or reject its possibilities for new innovations? Emotions can contribute to excitement about the impact of science on society and culture, or they can be linked with skepticism or even doubt and denial of scientific consensus. Kenisha's experience helps identify the problems this view of science understanding presents and gives new directions for promoting greater scientific literacy.

How the Public Got "Plutoed"

When Pluto was demoted to dwarf planetary status in 2006, it set off a firestorm of controversy. Overnight, our solar system went from nine to eight planets, and a new definition of "planet" was adopted by the International Astronomical Union (IAU). The reaction was overwhelmingly negative in the general public and among young children, more negative than perhaps most members of the IAU suspected it might be. In reaction to the public outcry, Neil DeGrasse Tyson wrote a book about the IAU's decision and the public reaction called *The Pluto File: The Rise and Fall of America's Favorite Planet*.[1] The term "plutoed" entered the vernacular, meaning to be demoted or devalued.[2] In that book he describes receiving "endless hate mail from third graders"[3] for his role in advocating for Pluto's reclassification.

Gale and her doctoral student Suzanne Broughton[4] decided to teach 5th- and 6th-grade students about the reasons the IAU had decided to change Pluto's status and see if that would help students understand and accept the new scientific definition. At the same time, they wanted to check on how students were feeling about it. They presented the elementary school students with a text designed to refute students' misconceptions about the reclassification and engage some of the students in more in-depth discussions.

Emotions about the reclassification decision were strong, and students' attitudes were decidedly negative. Exclaimed one exasperated 12-year-old, "I was mad and frustrated. . . . I just thought it would be a planet because my whole life I knew it as being a planet and now that it's not, it just doesn't seem the same."[5] Other emotions expressed by the students were more wistful, as reflected in this student's comment: "I felt kind of sad because I wanted it to be a planet."[6] One student described his reaction as "pretty surprised because everyone used to think that was a really small planet.

And I thought it was a planet too . . . so I was pretty surprised."[7] Surprise is typically considered a neutral emotion (neither positive nor negative), but in this study surprise was more closely associated with negative emotions (such as sadness) than is typically seen, indicating that the news about Pluto's demoted status was not a welcomed surprise. After learning about the (IAU's) rationale for reclassification, some students described how they were open to the new definition of "planet," if it applied only to planets not yet discovered. It was as if they wanted Pluto's status to be "grandfathered" in so that it could forever remain one of the original nine planets in our solar system.

Students expressed a variety of emotions about the scientific enterprise it-self, such as surprise to hear that something in science that seemed so basic, like the number of planets, can change from nine to eight. One student re-flected a sense of frustration with the changing nature of scientific know-ledge when she noted, "It's been a planet for a long time, so why not just keep it a planet?"[8] Another student expressed doubt about the decision and won-dered whether the scientific consensus might yet change again, noting, "I'm still surprised because I don't know, they might change it back."[9]

The more negative the students' emotions, the greater the resistance they expressed to accepting the IAU's reclassification. However, even emotions about knowledge that may seem to be negative, such as confusion, can moti-vate us to learn more about a topic, playing a positive role in science thinking and reasoning.[10] While the students in this study did come to a better under-standing and their negative attitudes softened a bit, they were still upset. As one student summed it up, "Pluto has been a planet my whole life!"[11]

This study illustrates several points about the challenges of understanding how emotions and attitudes influence science education. Sometimes it is hard to know what students are thinking or feeling about a topic before they learn about it. Teachers might suspect certain topics are naturally "hot button" issues, but they can be caught off guard by students' passions about certain topics. For example, as an assistant professor in Utah, Gale conducted a reading comprehension study using a text on biological evolution. She was surprised by the level of resistance to the science of evolution, the raw emotions triggered by the reading, and how much students associated their identity with their views about science. Students responded with such statements as, "if I were to believe we are related to animals, I would have no reason to go on living." Viewing humans' relation to other living things as wonderous and awe-inspiring, Gale was unprepared for their reactions.

When educators are prepared for strong negative reactions, they can design instructional activities to help students manage their negative emotions and understand the science by explaining how scientists think and why they think that way.

Beyond the Cold View of Science

As the Pluto case illustrates, individuals experience a variety of emotions when encountering scientific information. There may be awe when considering the vastness of the universe or fear when experiencing how climate change might cause sea levels to rise to dangerous heights in a coastal hometown. Individuals can be curious about why scientists think that COVID-19 may have come from bats in China[12] or confused about how light cannot escape the gravitational pull of a black hole. Your aunt might be angry when she hears that scientists think that humans are related to animals, and yet your cousin might feel a deep sense of connectedness to the rest of the biological world.

Thoughts Are Linked to Emotions

Emotions are often portrayed as something to keep under control. Adults are told to not let their emotions get the better of them and not to show their true feelings in public, or, as the British saying goes, "Keep calm and carry on." Consider how emotions are sometimes portrayed in film and television: to be "emotional" is to be "irrational." This suggests that when emotions are involved, the result is thoughts and actions that are the opposite of what we might think of as the rational, scientific reasoning process.

Traditionally, emotions were considered to be feelings quite apart from what one might think about those feelings. More recently, psychologists and neuroscientists consider emotions to be the sense individuals make of their bodily sensations in a particular context.[13] In other words, you might feel a flutter in your stomach right before giving a public speech, and you would interpret that as anxiety, while a similar physiological experience just before watching a favorite sporting event might be interpreted as excitement.

Work by Mary Helen Immordino-Yang[14] and other neuroscientists[15] studying emotions indicates that rather than some "extra-rational" entity that takes away from the ability to think and act clearly, emotions are instead the very platform for reasoning. The new view of human cognition has gone from "I think, therefore I am," to "I feel, therefore I learn."[16] Immordino-Yang's work shows that the thought processes individuals need to engage in effective scientific reasoning, such as paying attention, forming memories, and making decisions, rest on how individuals make sense of emotions, what she calls "emotional thought."

Emotions are not just an individual or biological process but are culturally embedded as well.[17] How individuals experience emotions, label them, and learn to recognize them on each other's faces are all deeply embedded in cultural experiences.[18] You may have noticed this when interacting with individuals from different cultures who may be more or less expressive of their emotions than you are or who may react with different emotions than you do to the same circumstance. When and where we learn about the social world influence how we interpret and express emotions.

Emotions Play Different Roles in Thinking About Science

Emotions are experienced when reading or learning about science and they also influence understanding. Learning about fuel cells as a possible way to generate clean energy can inspire feelings of hopefulness, which may prompt reflection and deeper thinking about the science of alternative energy. It may also create a desire to learn more about solar and wind energy, which may promote even greater learning about energy. In contrast, hearing that significant sections of a city you love, like Houston or Miami, have suffered severe damage due to "century storms" that seem to be coming more and more frequently may inspire hopelessness, which could create resistance to careful consideration of the threat to coastal communities. It may even discourage further engagement with the topic and a tendency to swipe past, rather than click on, articles describing coastal damage. These emotions may also play out in just the opposite manner for someone else reading about fuel cells who may feel threatened by the prospects of alternative energy and its impact on the economy or an individual excited to learn more about how sea walls might be constructed to prevent further destruction of seaside communities.

Fear may have played a significant role in how individuals perceived the threat of COVID-19, leading them to be drawn to articles or cable news shows that make the claim that this is just another flu or alternatively to learn more about how masks and hand sanitizer minimize one's risk. Researchers showed that more than other factors, such as personality characteristics or political affiliation, the primary predictor of the degree to which individuals adopted public health recommendations such as handwashing and keeping socially distant was fear of contracting the virus.[19]

Some emotions help us understand science, while others block our understanding. Emotions can be thought of in two ways: the first is whether they are positive or negative, what psychologists call "emotional valence," and the second is intensity, or how strongly the emotions are experienced. Positive emotions, such as enjoyment and interest, tend to have a positive impact on understanding and learning. When you enjoy a new experience, the outcomes (in this case, how much you learn) are likely to be more positive. There is a good reason for this: emotions such as enjoyment are linked with effective learning strategies. When you enjoy reading about whether humans could live on Mars or about the structure of a cell, you are more likely to employ active strategies such as trying to relate the content to something you already know, thinking critically about it, or diagramming your understanding. Such study strategies are more likely to make the information stick.[20] When you are not enjoying what you are reading about online or learning about in a classroom or if you are bored or angry, you shut down those active and engaging study approaches, perhaps clicking to a new article online or engaging in off-task activities such as thinking about what you are going to have for lunch. In these cases, emotions lead individuals to disengage from thinking or reasoning, leading to less learning.

Specific content can also spark emotions, such as a feeling about mathematics in general (perhaps anxious) or about having to solve word problems in particular (maybe scared), and these can be felt with more or less intensity. Such responses influence how people react to different science topics. In our own research we have seen that certain topics tend to engage emotions more intensely than others. Photosynthesis or seasonal change do not usually prompt strong emotional reactions, but human evolution and climate change often do. Sometimes these emotions are positive, such as the interest one feels when learning that wind turbines can provide a clean energy source. But often these emotions can be negative, such as feeling angry or threatened by learning about the impact of climate change. This is why some science topics are not particularly difficult to teach or learn about, while others are more challenging.

In fact, climate anxiety, or "eco-anxiety," is how symptoms such as panic attacks, sleeplessness, irritability, feelings of guilt, or obsessive worry about climate change have been described. Such symptoms are an increasing concern of psychologists.[21] Eco-anxiety is increasingly prevalent among youth, who view themselves as the prime target of future climate impacts. These strong emotions have also led to more positive outcomes such as the wave of youth climate activism.[22] Emotions around science topics can be strong and influential, for either a positive or a negative outcome.

What Are Attitudes?

Are you for or against a carbon tax to mitigate climate change? Are you for or against labeling genetically modified organisms consumed for food (GMOs)? What about nuclear energy? Vaccines? Colonizing Mars? What is your attitude toward these scientifically based initiatives? An attitude is an individual's evaluation of a person, object, or entity.[23] Similar to emotions, attitudes are described as having a "valence" (e.g., pro or con, favor or disfavor, like or dislike).

Attitudes toward science are general evaluations about science. These judgments include how you think about something (the cognitive component), how you act toward it (the behavioral component), and how you feel about it (the affective component). Perhaps you *think* that science is an economic engine that creates jobs in the new knowledge economy and provides solutions to social and environmental challenges. Your favorable attitude toward science leads you to *take actions* such as participating in the Science March in your community. And finally, you may *feel* angry to learn of a proposed rescinding of policies on the use of energy-saving lightbulbs in spite of strong support for their use from environmental scientists.

Attitudes toward science can play out in a variety of ways in individuals' daily lives. Consider Maria, who has a positive attitude toward science. Maria thinks science provides many societal benefits such as medical advances and economic innovations. Maria enjoys watching the latest NOVA special on television and looks forward to her daughter's science fair competition. But, of course, attitudes can also be less favorable. Patrick struggled with science classes in high school because he never saw the relevance of what he learned in these classes to his daily life. Patrick avoids programs that are science-oriented and may not search online for the scientific explanation of an upcoming solar eclipse. Patrick may even avoid learning about science altogether. He may also be suspicious of the motives of scientists who make claims about human impact on the climate, as we have seen in our research,[24] or skeptical about the risks associated with COVID-19. He may wonder why scientists engage in what he considers ethically questionable research practices, such as cloning or stem cell research.

Attitudes are usually aligned with interests. This results in a rich getter richer and poor get poorer outcome. In other words, those interested in science seek out opportunities to learn more about it, which results in greater interest and, often, more positive attitudes, whereas those who are disinterested may tune out whenever a science topic comes up, and their negative attitude may not have an opportunity for change. The alignment isn't perfect because someone could have a neutral attitude toward science (neither positive nor negative) and simply not be interested in science.

In contrast to attitudes toward science, philosopher Lee McIntyre described what he calls a "scientific attitude."[25] An individual, either a scientist or a member of the public, adopts a scientific attitude when they care about evidence and are willing to change their thinking in light of new evidence.[26] In contrast to a pro or con valence, a scientific attitude as McIntyre describes it is a particular approach toward consideration of scientific information. In other words, caring about evidence means not dismissing new or contradictory findings out of hand. Those with a scientific attitude are willing to seek out evidence and truly subject their ideas to the test using scientific practices.

Emotions and attitudes are intertwined with science acceptance as well as science skepticism and denial, and they direct and guide decision-making about critical scientific debates. When parents choose to homeschool their children rather than vaccinate them (in states where vaccinations are required for school attendance), strong emotions, such as anxiety about their children's health, are intertwined with attitudes toward the science on vaccination safety. When voters evaluate a ballot measure for labeling GMOs, their views about the science are linked to their attitudes for or against labeling.

Gale and her colleagues have explored attitudes toward GMOs and the link between attitudes and emotions.[27] Specifically, they found that in Southern California, where there is strong resistance to GMOs, many individuals had negative emotions about GMOs that were associated with negative attitudes about GMOs. If individuals were angry or fearful about GMOs, they also tended to report that they were against GMOs and would not eat them. They also found that these individuals had many misconceptions about GMOs, such as that GMOs involve cloning. When they read a text designed to refute their misconceptions, their emotions became less negative, as did their attitudes. Emotions and attitudes are closely linked and tend to shift in the same direction, either both becoming more positive or both more negative over time or with new information.

Although research suggests that K–12 students can hold unfavorable attitudes toward science classes, there are many students who are exceptions.[28] Attitudes can vary, related to gender, race, and ethnicity. For example, males typically show more positive attitudes toward learning and doing science than females.[29] However, this gap varies depending on the discipline, with larger interest and attitude gaps between males and females evident in physics and computer science and even some advantage for females over males in enrollment and interest in biology-related majors and careers. Whether or not a particular individual chooses a science career depends not only on their attitudes toward science but also on other important factors such as their sense of belonging in the field. Women and members of underrepresented groups are often made to feel as though they do not belong in the hard sciences.[30] When individuals are made to feel like outsiders, they have a much harder time succeeding in their chosen major or career.

Emotions and Attitudes in the Public Learning of Science

In the age of online gaming and social media, one might think that zoos, museums, and national parks are going the way of the dodo bird. This is far from the case. Tens of millions of people visit zoos, national parks, and museums annually in the United States. The San Diego Zoo has 3.5 million visitors a year. In the United States, national parks are so well attended they are becoming "loved to death," with hordes of visitors causing Zion National Park to restrict personal vehicles and employ buses for park entrance.[31] Science museums are no exception. The California Science Center receives over 1.5 million visitors annually. Even during the shutdown associated with the coronavirus in 2020, museums offered online events and exhibit tours and may have gained even more followers. These informal learning spaces provide valuable opportunities to learn through lectures, films, programming for students, families, and community members. These opportunities result in meaningful learning gains.[32]

Museum exhibit designers know the benefits of drawing on visitors' emotions as they design exhibits to help the public learn about science. The plight of the polar bear, given the impacts of climate change on their shrinking environment, has drawn on the heartstrings of many zoo visitors. A zoo visit can engender a feeling of personal connection to the impacted animal and that could make concern about climate change more personally relevant.[33] This is what led Alejandro Grajal, senior vice president of the Chicago Zoological Society, and his colleagues to see an opportunity to draw on the public's emotional response to the polar bear to promote understanding of climate change impacts. He noted, "People don't make their daily decisions based on scientific facts. There is an important emotional and psychological component to learning and decision making, so we're trying to understand how those processes work in the particular case of zoo visitors."[34] The challenge is how to educate the public within the political climate that engenders as many emotions and attitudes as the science does.

A feeling of connection to nature can increase the personal relevance of the zoo or museum visitors' experiences and encourage concern about climate change within a supportive social context. When Grajal teamed up with Susan Clayton and others to research this issue, they found that the zoo visitors they surveyed who showed a high degree of emotional connection to the animals in the zoo were likely to say that citizens should take actions to mitigate climate change.[35] They were aware that connecting the plight of the polar bear to climate change might put off some visitors who view climate change as a political issue. While highlighting the effects of decreasing polar ice on the bear's habitat, they also trained museum staff to be prepared to handle angry or skeptical questions from visitors who might be offended by the exhibit.

Emotions and Attitudes in the Learning of Science in School

In classrooms and at home, teachers and parents know that students do not check their emotions at the door when learning about science. Nor should they; emotions serve as a mechanism by which students make learning meaningful.[36] Students can experience anxiety over an upcoming exam or project deadline, confusion when trying to solve a quadratic equation, or enjoyment when succeeding on a challenging homework assignment.[37] Reinhard Pekrun and his colleagues have shown that these *academic emotions* have a significant impact on learning about science as well. A student who is enjoying conducting an experiment in a chemistry lab is more likely to focus on the activity, leading to positive results such as conducting the experiment correctly (and safely) and learning some chemistry.[38] In contrast, negative emotions, such as anxiety over an upcoming test, can drain emotional resources and make focusing on reviewing for that exam all the more difficult.

Emotions influence learning in other ways beyond just focusing attention toward or away from learning tasks and activities. They can affect motivation, interest, memory, and the type of study strategies employed. When students are motivated to engage deeply with content that they enjoy, they are likely to go the extra mile to learn it. Students are much more likely to do the heavy lifting needed for critical thinking and strategic problem-solving when they are enjoying a task than when they are angry about doing it or bored by it.

Students in classrooms also experience emotions about a specific topic, such as frustration or excitement when trying to program a robot, fear or anxiety when learning about GMOs,[39] or sadness and anger when learning about Pluto's reclassification to a dwarf planet.[40] Students' emotions might shift as the classes in the school day shift from math to history to science. These emotions might even shift within a class period. Trevor might really enjoy biology class and find learning about how cells react to antibodies really interesting as he remembered how his dad recovered from an infection he got when traveling. But if the topic shifts to the role of evolution in the development of superbugs that are resistant to antibiotics, the emotions might shift to concern over his own health or anger that the topic has shifted to evolution, something his faith has led him to question.

Students also have emotions about knowledge itself and about the learning process. For example, students can be curious about how scientists know that GMOs are safe to eat, or they can be frustrated by their teacher's choice to give a lecture on kinetic energy rather than allowing students the opportunity to explore the topic in a lab or field experience. These emotions may not be focused on the topic per se. A common situation in a science classroom that is likely to arouse emotions about knowledge and learning is called "cognitive incongruity."[41] This occurs when students' ideas are not aligned with what they are learning or when they hear or read conflicting information about the same topic (e.g., one website says social distancing reduces the spread of viruses, and another says that the fastest way to reach herd immunity is to allow individuals to interact). This may spark confusion, surprise, or curiosity. This situation is not uncommon in a classroom when students come to the study of science with many ideas that may conflict with what they are being taught. For example, they may have heard from their parents that vaccines cause autism, but their science teacher may tell them that vaccines are safe and critically important. Emotions about a specific topic or a learning situation that are somewhat negative (like anger or frustration) may motivate students to resolve their feelings by reconciling different points of view; and the outcome could be positive for learning science,[42] but often negative emotions lead to less learning.[43]

The students in Gale's study who heard about Pluto's diminished status experienced a variety of emotions including anger and surprise. Students' emotions were related both to their learning about the new definition of a planet put forth by the IAU and to their attitudes about that decision. Students with more positive emotions showed that they understood the new definition of a planet better than those who were really angry or frustrated by Pluto's ostensible downgrade. Students who had negative emotions about Pluto's new status also reported that they were against the IAU's decision. Indeed, emotions and attitudes about science topics are inextricably linked, whether they are experienced by adults seeking out information about a science topic online or by elementary school students learning about the solar system.

What Can We Do?

Understanding science involves the full range of human emotions, including joy, amazement, surprise, and confusion as well as anxiety, anger, fear, and hopelessness. These emotions are present when individuals are learning in the science classroom and when they explore topics online. Understanding science is never a "cold" and rational process, but rather, as research clearly shows, emotions, attitudes, and beliefs are deeply intertwined in thinking and reasoning about science, as they are with all of the human experience. The problem is that some emotions support thinking and reasoning and others pull individuals away from engaging with tough topics that require full attention and consideration.

What Can Individuals Do?

The internet has put the contents of the world's libraries just a few clicks away for those who have access. This available mass of information, both scientific and fraudulent alike, has the potential to evoke strong emotions and attitudes. Since it is not really possible, or necessarily helpful, to check your emotions before engaging with the latest science information found online, recognizing the role they play in decision-making is vital if individuals are to evaluate information wisely for their health and well-being.

Many of us have seen the phenomenon of adults in online social groups who shut down or even "unfriend" those who bring up controversial topics, probably because they do not want to experience strong negative emotions in their social experiences. The same can happen if you are researching a science topic online. These strong emotions can lead to what psychologists call "foreclosure," shutting down the critical thinking process prematurely before one has the information needed to make an evidence-based decision.[44] For example, many individuals are avoiding specific foods based more on fads than on true medical risk factors.[45] A small percentage of people cannot tolerate gluten, yet gluten-free products are broadly popular. Some individuals may be missing out on easy ways to get key nutrients by avoiding foods that are fine for them. Recognizing how negative emotions and attitudes may be leading to rejection of scientific information without full evaluation is a first step. Be open to new ideas, and be sure to evaluate them critically.

What Can Educators Do?

Traditional views of well-behaved students with hands crossed sitting quietly in neat rows reading the textbook have been outdated for quite some time. Engaging classrooms today might seem chaotic to a teacher transported to the present from the early 1900s and the one-room schoolhouse. While this time-traveling teacher probably hoped to avoid expressions of emotions in the classroom, exemplary teachers today expect that students will get excited, frustrated, anxious, hopeful, and sad at different points during their school day. Rather than avoiding emotions, the new perspective on emotions as a platform for learning suggests that teachers should welcome emotions into their classroom and begin to realize the crucial role they play in meaningful learning.

Students learn best when emotions are positive, but that doesn't mean that learning can always be fun. There are many topics that are inherently sad or disturbing in science, such as animal extinction. Students can have unique or unexpected emotional reactions, and teachers must be prepared to provide a safe space for students' potentially varied or unexpected responses to lessons or activities. Positive emotions like enjoyment must be centered on the content to have an impact. In Gale's research, she has found that reducing negative emotions about what students are learning is a good way to increase the likelihood of positive learning outcomes. Enjoying what you are learning is best, but not hating it is a good first step.

What Can Those in Informal Learning Environments Do?

From science museums to summer camps to science programing afterschool to television shows to online resources, there are countless opportunities for students and adults to experience positive emotions and attitudes while learning about science outside of academic settings. Museums may have an edge over classrooms because they are opt-in. Although they are responsible to their governing boards and donors, they are freer than public schools to present controversial topics without worrying about interference from nervous school boards or complaints from angry parents. After-school programs that instill positive emotions and attitudes toward science are available in countless communities and not just those that are well-resourced. Citizen science programs, although not without their critiques, have the possibility of engaging adults in learning about (and even conducting) science in their own backyards.

What Can Science Communicators Do?

There is much that can be learned from research on emotions and attitudes about how to communicate science effectively. Scientists and science writers have become increasingly aware that science news is often met with skepticism and strong emotions and attitudes. Awareness is a first step. The next step is to acknowledge those emotions and attitudes when communicating. Stephen J. Gould, a brilliant science communicator, liked to remind his readers that "nature contains no moral messages"[46] when he shared disturbing stories about the animal world such as parasites that eat their host while keeping it alive. He was keenly aware that readers would likely impose their values and emotions on the animal world and shut down before understanding the point of his essay.

To the extent possible, science communicators can be most effective when evoking positive and reducing negative emotions and attitudes in conveying scientific information. Sharing action steps that readers can take to mitigate their impact on the environment or improve their health is a positive message that can help readers move from negative emotions like frustration and sadness to more positive feelings of hopefulness and curiosity. Promoting awe and wonder at the potential of scientific discoveries can help balance the negative emotions and attitudes spurred by concerns about dangerous negative potential impacts.

Conclusions

Emotions impact acceptance or rejection of the scientific information in and outside the classroom experience. Research clearly shows that emotions and attitudes are deeply intertwined in thinking and reasoning about science, as they are with all human experiences.

Notes

1. Neil deGrasse Tyson, *The Pluto Files: The Rise and Fall of America's Favorite Planet* (New York: WW Norton, 2009).
2. EarthSky, "How Did Pluto Become a Dwarf Planet?," *Human World* (blog), August 24, 2019, https://earthsky.org/human-world/pluto-dwarf-planet-august-24-2006.
3. Tyson, *The Pluto*.
4. Now Suzanne Broughton Jones, PhD.
5. Suzanne H. Broughton, "The Pluto Debate: Influence of Emotions on Belief, Attitude, and Knowledge Change" (PhD diss., University of Nevada, Las Vegas, 2008).
6. Broughton, "The Pluto Debate."
7. Broughton, "The Pluto Debate."
8. Broughton, "The Pluto Debate."
9. Broughton, "The Pluto Debate."
10. Sidney D'Mello et al., "Confusion Can Be Beneficial for Learning," *Learning and Instruction* 29 (2014).

11. Suzanne H. Broughton, Gale M. Sinatra, and E. Michael Nussbaum, "'Pluto Has Been a Planet My Whole Life!' Emotions, Attitudes, and Conceptual Change in Elementary Students' Learning About Pluto's Reclassification," *Research in Science Education* 43 (2013).

12. Muhammad Adnan Shereen et al., "COVID-19 Infection: Origin, Transmission, and Characteristics of Human Coronaviruses," *Journal of Advanced Research* 24 (2020).

13. Lisa Feldman Barrett, *How Emotions Are Made: The Secret Life of the Brain* (New York: Houghton Mifflin Harcourt, 2017).

14. Mary Helen Immordino-Yang and Antonio Damasio, "We Feel, Therefore We Learn: The Relevance of Affective and Social Neuroscience to Education," *Mind, Brain, and Education* 1, no. 1 (2007).

15. Barrett, *How Emotions Are Made*.

16. Immordino-Yang and Damasio, "We Feel, Therefore We Learn," 3.

17. Mary Helen Immordino-Yang and Xiao-Fei Yang, "Cultural Differences in the Neural Correlates of Social–Emotional Feelings: An Interdisciplinary, Developmental Perspective," *Current Opinion in Psychology* 17 (2017).

18. Mary Helen Immordino-Yang, *Emotions, Learning, and the Brain: Exploring the Educational Implications of Affective Neuroscience*, Norton Series on the Social Neuroscience of Education (New York: WW Norton, 2015).

19. Craig A. Harper et al., "Functional Fear Predicts Public Health Compliance in the Covid-19 Pandemic," *International Journal of Mental Health and Addiction*, published ahead of print April 27, 2020, doi: 10.1007/s11469-020-00281-5.

20. Peter C. Brown, Henry L. Roediger III, and Mark A. McDaniel, *Make It Stick* (Cambridge, MA: Harvard University Press, 2014).

21. Thomas J. Doherty and Susan Clayton, "The Psychological Impacts of Global Climate Change," *American Psychologist* 66, no. 4 (2011).

22. Albert Bandura and Lynne Cherry, "Enlisting the Power of Youth for Climate Change," *American Psychologist* 75, no. 7 (2020).

23. Alice H. Eagly and Shelly Chaiken, *The Psychology of Attitudes* (Fort Worth, TX: Harcourt Brace, 1993).

24. Doug Lombardi and Gale M. Sinatra, "Emotions When Teaching About Human-Induced Climate Change," *International Journal of Science Education* 35 (2013).

25. Lee McIntyre, *The Scientific Attitude: Defending Science from Denial, Fraud, and Pseudoscience* (Cambridge, MA: MIT Press, 2019).

26. McIntyre, *The Scientific Attitude*, 48.

27. Benjamin C. Heddy et al., "Modifying Knowledge, Emotions, and Attitudes Regarding Genetically Modified Foods," *Journal of Experimental Education* 85, no. 3 (2017).

28. R. Tytler and J. Osborne, "Student Attitudes and Aspirations Towards Science," in *Second International Handbook of Science Education*, ed. Barry J. Fraser, Kenneth G. Tobin, and Campbell J. McRobbie (Dordrecht, The Netherlands: Springer, 2012).

29. Jennie S. Brotman and Felicia M. Moore, "Girls and Science: A Review of Four Themes in the Science Education Literature," *Journal of Research in Science Teaching* 45, no. 9 (2008); L. Raved and O. B. Z. Assaraf, "Attitudes Towards Science Learning Among

10th Grade Students: A Qualitative Look," *International Journal of Science Education* 33, no. 9 (2011); Brotman and Moore, "Girls and Science."

30. Ann Y. Kim, Gale M. Sinatra, and Viviane Seyranian, "Developing a Stem Identity Among Young Women: A Social Identity Perspective," *Review of Educational Research* 88, no. 4 (2018).

31. Christopher Ketcham, "The Future Is the Car-Free National Park," *The New* Republic, April 10, 2018.

32. National Research Council, *Learning Science in Informal Environments: People, Places, and Pursuits* (Washington, DC: National Academies Press, 2009).

33. Susan Clayton et al., "Connecting to Nature at the Zoo: Implications for Responding to Climate Change," *Environmental Education Research* 20, no. 4 (2014).

34. Lauren Morello, "Can Zoos Play a Role in Climate Change Education?," *Scientific American*, December 1, 2011.

35. Clayton et al., "Connecting to Nature at the Zoo."

36. Immordino-Yang, *Emotions, Learning, and the Brain.*

37. Reinhard Pekrun et al., "Academic Emotions in Students' Self-Regulated Learning and Achievement: A Program of Qualitative and Quantitative Research," *Educational Psychologist* 37, no. 2 (2002).

38. Jörg Meinhardt and Reinhard Pekrun, "Attentional Resource Allocation to Emotional Events: An ERP Study," *Cognition and Emotion* 17, no. 3 (2003).

39. Suzanne H. Broughton and Louis S. Nadelson, "Food for Thought: Pre-Service Teachers' Knowledge, Emotions, and Attitudes Toward Genetically Modified Foods" (paper presented at the annual meeting of the American Educational Research Association, National Conference, Vancouver, Canada 2012).

40. Broughton, Sinatra, and Nussbaum, " 'Pluto Has Been a Planet My Whole Life!' "

41. Reinhard Pekrun et al., *Emotions at School* (New York and London: Routledge, 2017).

42. D'Mello et al., "Confusion Can Be Beneficial for Learning."

43. Heddy et al., "Modifying Knowledge, Emotions, and Attitudes Regarding Genetically Modified Foods."

44. David L. Blustein and Susan D. Phillips, "Relation Between Ego Identity Statuses and Decision-Making Styles," *Journal of Counseling Psychology* 37, no. 2 (1990).

45. Aaron E. Carroll, "Relax, You Don't Need to 'Eat Clean," *New York Times*, November 4, 2017, https://www.nytimes.com/2017/11/04/opinion/sunday/relax-you-dont-need-to-eat-clean.html?_r=0.

46. Stephen Jay Gould, *Hen's Teeth and Horse's Toes: Further Reflections in Natural History* (New York: WW Norton, 2010), 42–43.

References

Bandura, Albert, and Lynne Cherry. "Enlisting the Power of Youth for Climate Change." *American Psychologist* 75, no. 7 (2020): 945–51.

Barrett, Lisa Feldman. *How Emotions Are Made: The Secret Life of the Brain.* New York: Houghton Mifflin Harcourt, 2017.

Blustein, David L., and Susan D. Phillips. "Relation Between Ego Identity Statuses and Decision-Making Styles." *Journal of Counseling Psychology* 37, no. 2 (1990): 160–68.

Brotman, Jennie S., and Felicia M. Moore. "Girls and Science: A Review of Four Themes in the Science Education Literature." *Journal of Research in Science Teaching* 45, no. 9 (2008): 971–1002.

Broughton, Suzanne H. "The Pluto Debate: Influence of Emotions on Belief, Attitude, and Knowledge Change." PhD diss., University of Nevada, Las Vegas, 2008.

Broughton, Suzanne H., and Louis S. Nadelson. "Food for Thought: Pre-Service Teachers' Knowledge, Emotions, and Attitudes Toward Genetically Modified Foods." Paper presented at the annual meeting of the American Educational Research Association, Vancouver, Canada, 2012.

Broughton, Suzanne H., Gale M. Sinatra, and E. Michael Nussbaum. "'Pluto Has Been a Planet My Whole Life!' Emotions, Attitudes, and Conceptual Change in Elementary Students' Learning About Pluto's Reclassification." *Research in Science Education* 43 (2013): 529–50.

Brown, Peter C., Henry L. Roediger III, and Mark A. McDaniel. *Make It Stick*. Cambridge, MA: Harvard University Press, 2014.

Carroll, Aaron E. "Relax, You Don't Need to 'Eat Clean.'" *New York Times*, November 4, 2017. https://www.nytimes.com/2017/11/04/opinion/sunday/relax-you-dont-need-to-eat-clean.html?_r=0 .

Clayton, Susan, Jerry Luebke, Carol Saunders, Jennifer Matiasek, and Alejandro Grajal. "Connecting to Nature at the Zoo: Implications for Responding to Climate Change." *Environmental Education Research* 20, no. 4 (2014): 460–75.

D'Mello, Sidney, Blair Lehman, Reinhard Pekrun, and Art Graesser. "Confusion Can Be Beneficial for Learning." *Learning and Instruction* 29 (2014): 153–70.

Doherty, Thomas J., and Susan Clayton. "The Psychological Impacts of Global Climate Change." *American Psychologist* 66, no. 4 (2011): 265–76.

Eagly, Alice H., and Shelly Chaiken. *The Psychology of Attitudes*. Fort Worth, TX: Harcourt Brace, 1993.

EarthSky. "How Did Pluto Become a Dwarf Planet?" *Human World* (blog), August 24, 2019. https://earthsky.org/human-world/pluto-dwarf-planet-august-24-2006.

Gould, Stephen Jay. *Hen's Teeth and Horse's Toes: Further Reflections in Natural History*. WW Norton, 2010.

Harper, Craig A., Liam P. Satchell, Dean Fido, and Robert D. Latzman. "Functional Fear Predicts Public Health Compliance in the COVID-19 Pandemic." *International Journal of Mental Health and Addiction*. Published ahead of print April 27, 2020. doi: 10.1007/s11469-020-00281-5.

Heddy, Benjamin C., Robert W. Danielson, Gale M. Sinatra, and Jesse Graham. "Modifying Knowledge, Emotions, and Attitudes Regarding Genetically Modified Foods." *Journal of Experimental Education* 85, no. 3 (2017): 513–33.

Immordino-Yang, Mary Helen. *Emotions, Learning, and the Brain: Exploring the Educational Implications of Affective Neuroscience*. Norton Series on the Social Neuroscience of Education. New York: WW Norton, 2015.

Immordino-Yang, Mary Helen, and Antonio Damasio. "We Feel, Therefore We Learn: The Relevance of Affective and Social Neuroscience to Education." *Mind, Brain, and Education* 1, no. 1 (2007): 3–10.

Immordino-Yang, Mary Helen, and Xiao-Fei Yang. "Cultural Differences in the Neural Correlates of Social–Emotional Feelings: An Interdisciplinary, Developmental

Perspective." *Current Opinion in Psychology* 17 (2017): 34–40.Ketcham, Christopher. "The Future Is the Car-Free National Park." *The New Republic*, April 10, 2018.

Kim, Ann Y., Gale M. Sinatra, and Viviane Seyranian. "Developing a Stem Identity Among Young Women: A Social Identity Perspective." *Review of Educational Research* 88, no. 4 (2018): 589–625.

Lombardi, Doug, and Gale M. Sinatra. "Emotions When Teaching About Human-Induced Climate Change." *International Journal of Science Education* 35 (2013): 167–91.

McIntyre, Lee. *The Scientific Attitude: Defending Science from Denial, Fraud, and Pseudoscience.* Cambridge, MA: MIT Press, 2019.

Meinhardt, Jörg, and Reinhard Pekrun. "Attentional Resource Allocation to Emotional Events: An ERP Study." *Cognition and Emotion* 17, no. 3 (2003): 477–500.

Morello, Lauren. "Can Zoos Play a Role in Climate Change Education?" *Scientific American*, December 1, 2011.

National Research Council. *Learning Science in Informal Environments: People, Places, and Pursuits.* Washington, DC: National Academies Press, 2009.

Pekrun, Reinhard, Thomas Goetz, Wolfram Titz, and Raymond P. Perry. "Academic Emotions in Students' Self-Regulated Learning and Achievement: A Program of Qualitative and Quantitative Research." *Educational Psychologist* 37, no. 2 (2002): 91–106.

Pekrun, Reinhard, Krista R. Muis, Anne C. Frenzel, and Thomas Götz. *Emotions at School.* New York and London: Routledge, 2017.

Polikoff, Morgan, Q. Tien Le, Robert W. Danielson, Gale M. Sinatra, and Julie A. Marsh. "The Impact of Speedometry on Student Knowledge, Interest, and Emotions." *Journal of Research on Educational Effectiveness* 11, no. 2 (2018): 217–39.

Raved, L., and O. B. Z. Assaraf. "Attitudes Towards Science Learning Among 10th Grade Students: A Qualitative Look." *International Journal of Science Education* 33, no. 9 (2011): 1219–43.

Shereen, Muhammad Adnan, Suliman Khan, Abeer Kazmi, Nadia Bashir, and Rabeea Siddique. "COVID-19 Infection: Origin, Transmission, and Characteristics of Human Coronaviruses." *Journal of Advanced Research* 24 (2020): 91–98.

Sinatra, Gale M., Ananya Mukhopadhyay, Taylor N. Allbright, Julie A. Marsh, and Morgan S. Polikoff. "Speedometry: A Vehicle for Promoting Interest and Engagement Through Integrated STEM Instruction." *Journal of Educational Research* 110, no. 3 (2017): 308–16.

Tyson, Neil deGrasse. *The Pluto Files: The Rise and Fall of America's Favorite Planet.* New York: WW Norton, 2009.

Tytler, Russell, and Jonathan Osborne. "Student Attitudes and Aspirations Towards Science." In *Second International Handbook of Science Education*, edited by Barry J. Fraser, Kenneth G. Tobin and Campbell J. McRobbie, 597–625. Dordrecht, The Netherlands: Springer, 2012.

8

What Can We Do About Science Denial, Doubt, and Resistance?

Science denial, doubt, and resistance are pervasive and troubling. Numerous surveys show a large discrepancy between what the public accepts as scientifically valid and what scientists accept. Whether the topic is evolution, global warming, or genetically modified organisms, US citizens seem to have a poor awareness of what scientists know, as well as scientists' level of certainty about knowledge on such key issues.

A refusal or reluctance to accept settled science can impede solutions for many of the problems that face modern society—notably, serious threats of human-caused warming of Earth's climate and global pandemics. The goal of this book has been to provide the social and cultural context of science denial and the psychological explanations that underlie susceptibility to it. Improving the public understanding, acceptance, and valuing of science can move us all toward saving the planet and improving the health and wellness of all lives and communities.

As we have shown, science denial and doubt are not new. Galileo was convicted of heresy in 1633 for the unsettling claim that Earth revolved around the sun. The proposal that microorganisms or pathogens caused disease was also slow to gain acceptance over prevailing beliefs that the role of miasma, or bad air, was the precipitating agent—beliefs that led to a massive loss of life that could have been prevented. In spite of more recent decades of progress in scientific research that have dramatically increased the human life span and standards of living for many, resistance to scientific evidence persists and has been magnified in the current internet era. Information, misinformation, and disinformation are easily available and often indiscernible from one another. Remarkably, even "germ theory denial" can now flourish on the internet, appearing as justification for some forms of alternative medicine as well as opposition to vaccines, seemingly predicated on the belief that diet and health practices alone are the root of all illness. The coronavirus spurred an "infodemic,"[1] or massive spread of mis- and disinformation,

Science Denial. Gale M. Sinatra and Barbara K. Hofer, Oxford University Press. © Oxford University Press 2021.
DOI: 10.1093/oso/9780190944681.003.0008

about the causes and treatments of COVID-19. Science deniers can find not only support for their beliefs online but also an entire community of like-minded people available to help sustain their worldviews and to defend them when threatened.

Many have argued that the current era is a "post-truth" period,[2] where it has become increasingly difficult to assess what is actually well substantiated. As the algorithms of online life drive what one sees and social media posts reinforce what one already believes, individuals can find it increasingly difficult to know what is scientifically valid. The skills needed for evaluating information online and discerning what is truthful have not kept up with the rapid proliferation of online information.

Confusion about what is scientifically valid has been further blurred by journalistic practices advocating "balance," even when such presentations misleadingly distort what scientists know. Research is clear that this practice has a negative effect on human understanding and that balance can *become* bias.[3] Furthermore, such tendencies are ripe for exploitation by corporate entities that benefit from scientific doubt, whether about tobacco and cancer links or the role of carbon dioxide in climate change.[4]

Improved science education, while critically important, will not be a panacea for these problems. No individual can master the expertise on all the scientific topics that influence one's life or require decisions and action, whether at the ballot box or in the doctor's office. Science education can, however, better prepare individuals and citizens by addressing not only content knowledge but also the underlying assumptions of how scientists know what they know. This type of education, advocated by the Next Generation Science Standards (NGSS), can help bolster abilities to interpret scientific claims. Fostering a scientific attitude, described as an openness to seek new evidence and a willingness to change one's mind in light of new evidence,[5] is an equally important aspect of science education.

Regardless of educational backgrounds and training, we are *all* vulnerable to misunderstanding science, to doubting and resisting what is accepted knowledge, and to being hoodwinked by merchants of doubt. In the previous chapters, we reviewed five key psychological constructs that serve to explain our susceptibility: cognitive biases, epistemic cognition, social identity, motivated reasoning, and emotion.

Cognitive biases. Simply phrased, the mind takes mental shortcuts much of the time. These include *confirmation bias*—the process of seeking, interpreting, and recalling information that aligns with preexisting beliefs—as well as the *availability heuristic*—a predisposition to believe what is most readily available. Individuals are also prone to an illusion of understanding, misjudging how much they actually know. Countering these and other cognitive tendencies requires effortful, critical thinking, an employment of the reflective, deliberative mind.[6]

Epistemic cognition. Whether learning in a classroom, encountering new information online, or trying to resolve conflicting truth claims, individuals engage their epistemic cognition.[7] Every one of us has beliefs about what knowledge is and how we know—whether through authority, testimony, evidence, or some reasoned integration. These conceptions of knowledge and knowing develop and evolve in patterned ways, and they influence how one interprets scientific claims and how sources of knowledge are evaluated.[8] Those who see knowledge as dualistic—right or wrong, black or white—seek an absolute degree of certitude and have low tolerance for the ambiguity that arises as knowledge develops. A multiplistic view of knowledge, by contrast, can lead individuals to see all claims as merely a range of opinions, and therefore equally valid. Individuals who perceive knowledge with an evaluativist view, however, understand knowledge as tentative and evolving, evidence-based, with criteria for evaluating expertise and authority. Evaluativistic habits of mind enhance science understanding. Epistemic trust (e.g., in the scientific community, one's doctor, a pastor, a particular news source, or an educator) is also a critical component of epistemic cognition and a part of the public understanding of science.

Motivated reasoning. Humans are motivated reasoners. Although sometimes motivated by a desire for accuracy and understanding, they are often motivated by prior beliefs and biases that cloud or even distort their reasoning. A series of research studies indicate that given the same data to interpret, those who support the conclusions and those who do not are likely to interpret the results quite differently.[9] Such motivations can easily shape whether one accepts valid scientific conclusions.

Social identity. We are all tribalists, identifying with the groups to which we belong and those with whom we share core values. These social involvements create an influential aspect of our identities and might include such social units as our professional networks, groups of parents with similar attitudes, political parties, religious organizations, or any other such group. Group membership allows us to take shortcuts about what we think and believe, in harmony with what our groups avow. Going against these values can be difficult. A woman interviewed for a *New York Times* story on the anti-vaccine movement described her desperation for support from the Facebook group of moms who, like her, advocated home births, cloth diapers, and breastfeeding—but no vaccines. "I was so desperate for their support that I compromised by delaying the vaccine schedule, so I wouldn't get kicked out of the group."[10] Social identity and social values form one of the foundations of motivated reasoning.

Emotions. When learning and thinking about science, individuals bring a host of emotions and attitudes to their understanding. Both conducting science and learning about it can involve a range of human emotions, including joy, amazement, fascination, and surprise, as well as anxiety, anger, fear, and hopelessness. Negative emotions can cloud judgment and prevent deeper learning. Positive emotions, such as scientific curiosity and awe, can foster an openness to science learning.

These five psychological constructs form a useful framework for understanding how individuals interpret science and how they can approach the process in a constructive manner. They provide insights that can help improve both individual thinking and conversations with others.

Solutions: A Field Guide to Addressing Science Denial, Doubt, and Resistance

From our discussions with public audiences we have identified a number of concerns that individuals have about science denial, doubt, and resistance. They want to know how to improve their own understanding of science and how to be less susceptible to spurious claims. They want improved awareness of their own cognitive land mines that lead them to doubt or deny. They want to know how to enhance their online skills so that they can evaluate information more thoroughly and know what authorities to trust. They want to improve their quest for accurate information about science, health, and environmental topics.

Those with whom we have conversed about these issues also want to know how to better understand those who deny science. The want to be able to talk to others with opposing views, whether family members, neighbors, colleagues, clients, or politicians and public officials. Both "water cooler" conversations in the office and family gatherings can be difficult when discussions include deep disagreements over scientific topics such as climate change, the use of fossil fuels, fracking, pandemic guidelines, or vaccinations. Doctors tell us they want to know how to talk to patients to promote an evidence-based approach to treatment. Science educators want to know how to prepare individuals for a lifetime of seeking and identifying valid scientific information for questions of interest and importance, personally, professionally, and as citizens. Scientists want to communicate about their work in ways the general public can understand. Many policy makers want to ensure that they are making decisions based on the best available scientific evidence.

We include many suggestions for addressing science denial, doubt, and resistance at the end of each previous chapter. Here we summarize, synthesize, and expand upon these ideas.

What Individuals Can Do

Cultivate a scientific attitude and an appreciation for the value of science. The first step individuals can do sounds simple but takes some effort. Value evidence, scientific methodology, and the scientists who work as a community to keep science a self-correcting enterprise. Stay open to new evidence and new findings. Recognize that the tentativeness of scientific findings does not undermine their value, particularly when there is a preponderance of evidence. This is especially important when the science is rapidly emerging and shifting during a crisis such as a pandemic.

Monitor your own cognitive biases. Practice reflecting on your own tendencies to seek only confirming evidence, and check the impulse. Counter "confirmation bias" by expanding your options for new (and conflicting) information beyond what immediately supports your preexisting beliefs. Knowing when it is worth the work and when it is not is half the battle. If this is a critically important issue to you, take the time to explore more deeply, keeping an open mind. Take time to evaluate the evidence presented. None of us can think effortfully all the time, nor do we want to, so become tactical about what is important to you.

Keep in mind that we can easily be persuaded when we reason with "confidence by coherence." It may seem that mothers who share your values about breastfeeding also would know best about vaccinations, but do your own research. Beware the "illusion of understanding" as it is easy to become overly confident about what we think we know. Watch your tendency to be persuaded by the power of a good anecdote, and seek scientific evidence to confirm or deny the message of the story you heard.

Engage in critical thinking. Critical thinking is clear, rational, analytical, and informed by evidence and involves analysis, interpretation, and judicious evaluation. Not surprisingly, it is relatively easier when background information and prior knowledge are high. Good critical thinkers know when they have adequate knowledge to effectively evaluate evidence and when they do not and thus need to rely on others' expertise about the topic. When you take the latter path, recognize that you need to focus on evaluating the source of information and its reliability rather than the information itself.[11] Such skills can be learned, developed, and refined.

Improve your ability to search and evaluate scientific claims and their sources. A basic framework involves learning to 1) identify the motives of articles or websites, 2) identify tone or bias, 3) be skeptical of sources and develop tools to check them, and 4) be aware that algorithms have targeted what appears in any search.[12] Learn to evaluate the results of your online searches the way a fact-checker would, reading laterally and not vertically, across sites, to get a wider, more impartial view.[13] Those who operate like fact-checkers also spend more time sorting through results, are slower to reach conclusions, and are most accurate in their assessment of the integrity of sources. Should you wish to give yourself a crash course in improving your skills, most universities have websites that teach information literacy to their students and can often be accessed by the public.

Advance your algorithmic literacy. Everything you do online is a function of algorithms, the computer-coded instructions that determine what you see in your social media feeds, with whom you connect on a dating site, what results you get in any web search, what purchases are recommended for you. Algorithms are the basis of artificial intelligence and support facial recognition, voice commands, and driverless cars; and they are replacing humans in a wide variety of roles. They are also reflective of the biases of programmers and data sets.[14] The inevitable progression toward greater use of algorithms as aspects of everyday life requires awareness of how they function, how they might be biased, and how to counteract such biases, particularly in your own search for scientific information. Learn how and where they operate and how they might be constraining what you see and hear.

Be alert to those who may be devaluing scientific knowledge and expertise. Be especially wary of those who discredit scientific knowledge in ways that support a corporate or political agenda. During the late spring of 2020, politicians who supported early reopening of businesses and reversal of stay-at-home policies designed to protect people from getting COVID-19 were motivated by their support for economic revitalization but were not always putting public health advice first. Be cautious about attempts to cast known truths as unsettled ones. Current claims about a "post-truth" society can easily numb one to the need to stay vigilant in asserting the basic principles involved in expecting and respecting truthfulness.

Know the role of your emotions. Be aware of your own hot-button topics, and recognize the role your emotions play in engaging with information on these issues. This will help you avoid foreclosure, leaping to a conclusion too quickly before getting all the information you need to make a thoughtful decision. Employ your "scientific attitude," remaining open to new ideas, even if you are emotionally attached to prior ideas; and then be sure to employ critical evaluation of the new ideas. Seek evidence. Be equally attentive to positive emotions that enhance science understanding, such as scientific curiosity, scientific awe. Nurture them.

Examine the motivations in your reasoning. Before you seek information on a topic, ask yourself what it is you want to believe, and then remain open to new, perhaps contradictory findings and new evidence. Seek the best available evidence, and assess its credibility and the sources. If you're searching online, consider the search terms you employ and whether they bias the search toward a particular conclusion. If you're reading graphs or an explanation of scientific findings that run counter to what you are motivated to believe, try to stay open-minded. Explain the findings to yourself, and ask if you would find them acceptable if they were about another topic, one with less emotional impact. Be as thoroughly skeptical of findings that support your position as those that don't.

Be willing to consider different points of view. Open yourself to scientific findings that run counter to your groups' beliefs. When close friends and affinity group members advocate a particular stance that you might question, ask for their sources of evidence, and dig deep to assess their validity. Share your doubts with someone you trust. Seek allies with different, scientifically based points of view, and find out how others have navigated such terrain. For example, if your mothers' group advocates delaying or avoiding vaccinations, try not to take that advice on faith; instead, explore why and what the research says, and then talk to others outside your group who have made different choices. Ask for their sources of knowledge, and search further on your own. Recognize that you're making a decision for the best interest of your child as well as your community.

Nurture the value of science and scientific thinking in others. Work with the next generation to foster scientific curiosity and a scientific attitude. Children love to ask why, for example; and you can help them figure out how you find the answers and how scientists know, and teach them about scientific processes. Teach science at home, through simple experiments, observations, explorations. Expose kids to outdoor activity and nature, and respond to their innate curiosity about the world by modeling your own fascination with the natural environment and your open attitude. Volunteer at a science museum, library, or science fair.

Vote for those who value, support, trust, and fund science. Help put people in decision-making roles who value science and who seek and provide evidence for their claims on scientific issues, whether that is climate change, fracking, marijuana legalization, vaccination exemptions, gun control safety, or alternative energy. Press candidates to answer questions about scientific issues, seek evidence, and hold accountable those who are elected and appointed to policy roles. Funding science fuels the economy. Public funding is less biased than corporate funders who may be tentative about scientific results that impact their bottom line. The open science movement means that government-funded research must be shared (no matter what the results) with the public who funded it.

What to Do in Communicating with Others

We all seem to know what doesn't work and how frustrating it can be to have conversations that lead to further polarization and that then may prevent us from engaging with those who have different views. We need to know how to find common ground as a basis for discussion and how to engage productively. The pressing scientific issues of the day require that we learn these skills.

Learn to really listen to others. Remember that you don't have to accept what someone else believes in order to learn about their reasoning. By genuinely understanding another person's thinking you are in a better position to offer alternatives. Listening to them may also encourage them to listen to you. Simply giving people more information won't help unless they are finding it worthwhile to pay attention and listen.

Enhance your own science knowledge. Many people erroneously believe that Earth is naturally getting warmer and that this is a normal process. To be persuasive on this issue, for example, educate yourself about the data, learn what the hockey stick graphs convey, know how increased carbon dioxide affects the planet. See the website "How Global Warming Works,"[15] for example, for explanatory armament. Similarly, pediatricians are front-line persuaders in advocating for childhood vaccinations and can address misinformation directly (e.g., the erroneous link to autism). Remember that topic rebuttal alone is insufficient.[16] Explain why the misinformation is wrong, and help them see why the alternative explanation is more plausible.

Listen for perspectives on knowledge that shape claims. If you're talking to someone who advocates a dualistic approach to knowledge and who argues that scientists just don't know for certain, talk about what scientists *do* know and about the confidence of their claims (e.g., the human causes of climate change), as well as how scientists value tentativeness—not to be confused with uncertainty. If the person reasons from a multiplistic view and sees all claims as opinions, talk about the value of evidence, where you got yours, how you evaluated it, and how to interpret it. Stay open to learning about your own perspectives on knowledge.

Seek to understand concerns and fears. Remember the power emotion plays in reasoning about hot scientific topics. On issues where you want to be persuasive, find out what the common worries are likely to be, and address them directly. Individuals who are reluctant to wear masks to stop the spread of the novel coronavirus or parents who are hesitant to vaccinate, for example, not only may have erroneous information; they also may be fearful about making a choice that inflicts potential harm on their children or elderly family members. Help assuage these concerns, even as you acknowledge that no medical intervention can be 100% safe.

Be aware that many disagreements are not about facts but about values. Often, people will disbelieve information that has scientific support if they are trying to protect a deeply held value.[17] They can also become frustrated when others seem insensitive to these values. Probe to understand what those values are.

Find points of common values. Climatologist Katharine Hayhoe, who claims both scientist and evangelical Christian as parts of her identity, cautions against thinking monolithically about others' beliefs. Instead of thinking about all the ways you and the person with opposing views differ, you can focus on what you have in common.[18] For example, in a conversation about climate change you might connect with a person who loves hiking or fishing by connecting that to caring about the environment or one whose love for their grandchildren can help them care about the future of the planet. Connect around common values.

Recognize that change is possible. A woman who changed her mind about vaccinations when her child was 9 (unphased earlier by the pleadings of her sister-in-law, a pediatrician) explained that her decision to vaccinate her daughter came from learning about herd immunity from a fellow parent whose child was too weak to be vaccinated. "I realized that I was putting the extremely slim chance that a vaccine could harm our child ahead of the reality that someone else's life could be endangered. Suddenly I felt I had been unspeakably selfish. . . . Just as there is scientific consensus that climate change is real, there is overwhelming scientific consensus that vaccines keep the whole community healthy. You are unlikely to find scientists who don't believe in climate change and you are unlikely to find scientists who don't vaccinate their children. As Neil deGrasse Tyson said, 'The good thing about science is that it's true whether or not you believe it.' "[19] Scientific attitudes develop over time. You might not have an immediate impact or even know if you've had one, but be aware that it's possible to be influential in guiding others toward an acceptance of scientific evidence.

What Educators Can Do

We have argued throughout the prior chapters that knowledge is not enough to address the challenges of public understanding of science. But we have also been clear that educators play a critical role in enhancing science knowledge in learners of all ages. Educators can increase their own appreciation of how students think and learn about science. With that understanding, they can improve their support of science learning. Educators also have the chance to influence students' appreciation of the contribution of science as well as its limitations. Students' abilities to think scientifically to solve problems in their lives and the lives of others in the community can be supported through carefully crafted instruction. Through instructional activities such as discussion, simulations, experiments, and instructional approaches designed to refute their misconceptions and provide an explanation of the scientific view, educators play a crucial role in promoting public understanding of science.

Enhance your own understanding of science learning. Students in formal and informal settings come to the science learning experience with ideas that may conflict with settled science. Although college students in an astronomy class may understand that Earth is a sphere, that doesn't mean they understand how the seasons change based on the tilt of Earth's axis. It is important to get a sense of what students think before instruction. If they have misconceptions, ignoring them will result in ineffective learning or pushback based on inaccurate knowledge. Once misconceptions are identified, educators can help students see the conflict between their view and the scientific view. Be aware that learning science involves not only learning about the content. It involves how students think and feel about the content.

Be aware of strong prior beliefs, attitudes, and identity. Imagine two groups of college students in a general education science course learning about genetic modifications to crops grown for food (GMOs), one in California and one in Nebraska. Students' prior beliefs, attitudes, and identity may vary depending on how a particular scientific issue is perceived in their community. California students might be concerned about GMO safety and want foods to be labeled so that they know what they are buying. Nebraska students might have grown up on a farm and know how crops are modified to resist pests or make them more drought-tolerant. They may be concerned that labeling

corn-based products as GMOs will put their family farms out of business. Similarly, students who may have felt the impact of superstorms
or the loss of coal jobs in their hometown may have different concerns
when learning about climate change impact. Educators can be keenly
aware of their local context but may need to appreciate that students
who just moved to their community or university students from other
states bring different attitudes to each topic. Helping students reflect
on prior assumptions and how they may influence their perspectives
will help educators as they discuss controversial science topics in their
classrooms.

Recognize students' emotions. Experienced teachers recognize and appreciate the role of emotions in learning. They know that emotions are not
something to be boxed out of the learning situation but rather are key
to successful learning. Students young and old can get excited, fasci-
nated, inspired, curious, anxious, confused, and frustrated throughout
learning about science. These emotions should not be ignored but recognized for what they are, fully integrated aspects of human learning.
With notable exceptions, positive emotions tend to work best for promoting learning. Confusion or even frustration may motivate a learner
to resolve those negative feelings and study more, but such negative
emotions when experienced repeatedly can also turn a student off from
learning more science.

Positive emotions can have a learning advantage, but not all topics are inherently enjoyable. Recognizing that students can have strong feelings
and providing a safe space to express them is important for successful
learning. Doomsday messaging about a pandemic or climate change
is more likely to induce stress than promote learning.[20] It's great when
students can enjoy what they are learning, but it is optimal to at least
not be depressed and anxious. When teaching about climate change
educators can help students direct their feelings toward taking pro-
active steps such as lowering their carbon footprint or helping their
school conserve energy, even as they are helped to understand that the
solutions are much larger than individual actions.

Promote instruction in digital science literacy. So much information about
science is gleaned from online sources that are unreliable or in some
cases designed to deceive. The result is a virtual grab bag of reliable, unreliable, and deceptive information about science online. Rather than
accepting the newest "scientifically based" health or wellness claim

uncritically, students can be taught to carefully consider information sources and critically examine the evidence behind science claims. Our existing digital literacy instruction has not kept pace with the blistering escalation of mis- and disinformation about science online. New and more sophisticated methods of sorting valid from invalid claims require a full arsenal of tools for evaluating sources of science information online. Help students make judgments about the scientific evidence they find online. Teachers may need to take extra steps to incorporate critical evaluation as part of their teaching about online information searches.[21] The bar must be raised for students to have any chance of discerning what is "fake science news" from valid sources of scientific information.

Foster an appreciation for how science works. Too often in the past, science was taught as a collection of facts to be mastered. Today, guided by the NGSS, K–12 educators are encouraged to teach fewer facts and more process.[22] The NGSS recommends that instruction promote an understanding of how scientific knowledge is produced and rigorously evaluated through a process of peer review. Students should understand that scientists are not encouraged to go along with the trends and endorse popular points of view but instead are trained to be critical of new ideas and demand supporting and compelling evidence.

Students at all levels should know what scientists can best provide: increasingly more accurate knowledge about the physical and natural world. They also need to know what scientists cannot provide: definitive answers to social, political, moral, and ethical dilemmas. Science provides theory and data as to "what is" but does not tell us "what should be." For example, science can predict when an area may become uninhabitable due to climate change–enhanced sea level rise, but science cannot tell us how we should accommodate those who are displaced from their homes.

Science learning takes place beyond classrooms. Informal learning environments such as zoos, museums, after-school programs, and summer internships play an important role in science education and science appreciation. Outreach to community members through field trips, citizen science programs, and free family nights offers tremendous opportunities to engage visitors in appreciating and even conducting science outside of formal classroom settings. These environments can also reveal how the science knowledge they are sharing is evaluated and engage visitors in their own citizen science projects and activities.

Foster scientific thinking in all students. Thinking scientifically entails adopting a scientific attitude,[23] an appreciation that evidence is important and the willingness to change views in light of new evidence. An individual may struggle to pick up scientific thinking without deliberate and thoughtfully crafted instruction designed to promote it. The disposition to think critically and analytically must be encouraged because our natural tendency is to make quick judgments and move on.[24] Individuals may be unwilling or unable to do the heavy lifting scientific thinking requires.

Educators in K–12 schools and in higher education have the time and space to promote scientific thinking through regularly questioning and examining evidence until these practices become habits of mind. Educators can provide students with the tools to think about how settled science (such as the shape of Earth) became settled by guiding them through the same process of evaluating evidence that scientists go through before coming to consensus.

Creative instruction can teach science as a method of reasoning that not only has the power to solve global problems (such as food scarcity and disease) but can also be applied to problem-solving in daily life. Today, with citizens sorted into like-minded communities both in their neighborhoods and in social media circles,[25] it is important that students question their assumptions, push back against their intuitions, and consider alternative points of view. With encouragement, each of us can slow down reflexive thought processes and ask, *What do I believe about climate change?* and *Why do I believe it? Have I really considered the evidence behind this belief?* Educators are uniquely positioned to prompt students to question their own reasoning as well as identify flawed thinking in others. It becomes easier to recognize the flaws in our own thinking and reasoning once we become practiced at identifying logical errors in the reasoning of others. It also takes a recognition that strong emotions and motivations may be impairing judgments.

Encourage students to apply science to solve real problems of genuine interest. Students should address science questions that connect to their experiences. Remember those experiments you did in high school chemistry? If not, you are not alone. If you engaged in science experiments that were not of interest to you, it's unlikely to have been a memorable experience. Actively engaging students of all ages in asking and answering questions they care about and that solve actual problems is much more effective.[26] The skills learned in scientific thinking and reasoning can and should be applied to relevant questions such as *How safe is vaping?* or *What's the best way to wash hands to prevent contagion?* or *How much sugar is in soft drinks?* Questions that cut across disciplines such as *If the school replaces grass with drought-tolerant plants, what are the trade-offs between water and cost savings and loss of open space for kids to play?* help students see science as useful.

What Science Communicators Can Do

Science communicators are on the front lines of the war on science. Science writers, journalists, essayists, bloggers, teachers writing science books for children, and textbook authors are all a part of the science communication community. Equally important are scientists who want to communicate about their own work to the public. Those who communicate about science provide a critical public service, particularly in the face of science denial. In 2017, the new administration took down the US Environmental Protection Agency's "Climate Change" web page.[27] Government scientists were busy backing up their data, and journalists were making the public aware of this effort to silence climate science communication. Science writers' efforts to inform the public about science and science policy then and now are an important part of the scientific enterprise.

Write about science. Science coverage in major traditional news outlets has dropped precipitously. According to Shawn Otto, author of *The War on Science*, from the 1990s to the middle of the first decade of the 2000s, "the number of major US newspapers with weekly science sections fell from ninety-five to thirty-four."[28] Otto decried how major newspapers like the *Washington Post* cut their science sections. We lament that the public has a poor understanding of the science issues facing their daily lives, but the dramatic lack of coverage of these issues in mainstream print media is definitely not helping. Better public understanding of science is dependent on broad public coverage of science and may also require a reinvigoration of science communication programs in

journalism schools. There are many ways to get stories out today beyond the traditional outlets, and more science journalism is critically
needed.

Many scientists are interested in joining the science communication space
to communicate their own research to the public. Some scientists are
moving away from the old model of exclusively publishing in academic
journals only their colleagues can read and understand and learning
to write for a broader public. Scientific conferences are increasingly
including sessions on science communication. Scientists realize they
must be part of the science communication effort, and they know they
cannot wait for other communicators to do that for them. They must
become science communicators themselves and push back on science doubt, resistance, and denial while advancing the public's understanding of science.

Know your audience. The first rule every good communicator knows
is, know your audience. Are you communicating to those who may
be skeptical about science or have reasons to resist the information?
Does your audience adhere to a well-known alternative point of
view? How much does your audience understand or misunderstand
about the topic? Remember there is always a chance you can reach a
member of your audience and help them come to a greater appreciation of science. Being aware of possible trigger points for your audience is the first step to developing a carefully crafted message that
debunks misconceptions and supports understanding. Knowing what
motivates your audience's skepticism or resistance will help you better
craft your message.

Know the likely misconceptions your audience may have. Before
sharing any science information with the public, consider what
misconceptions are out there. Stories on stem cells should not only
define what stem cells are but should refute the erroneous belief
that the only source is aborted fetuses.[29] Such prior beliefs should
be called out and carefully debunked (with explanations to avoid
backfire effects) before presenting more of the story. Where the science is unsettled, be clear about what is not known and how much
that matters to the story. For example, should a climate change story
spend more time on the uncertainty around the number of feet sea
level will rise (a number that depends on a variety of factors), or
should the story focus instead on the near certainty that a warming

climate exacerbates sea level rise? Leaving your audience in a state of confusion undermines understanding of scientific certainty. During the COVID-19 pandemic, recommendations regarding mask-wearing, whether contagion spreads from packages, and the pros and cons of keeping schools open all shifted. Be sure to communicate not only when scientific recommendations change but why they changed. We recommend focusing on what is known and carefully contextualizing the degree of uncertainty about what is still unknown.

"Both sides" is for opinions, not science. If 98% of climate scientists accept humans are affecting the climate, then journalists are at fault when they create the illusion of a debate by interviewing one who accepts and one who doesn't, in some misguided attempt at fairness. Individuals may then naively conclude that the issue has not yet been settled. Nor do flat-earthers deserve equal time devoted to their view alongside scientific evidence of Earth's shape. Repeating misconceptions or debunked theories can reinforce them or introduce them to a new audience.[30]

News is what is "new," but be careful to not overemphasize stories about fad cures or treatments (unless the goal is to debunk them) based on single studies as the evidence is insufficient for members of the public to take well-supported and reasoned actions. If you want to communicate about a possible groundbreaking treatment with modest evidence, be careful to contextualize the results, state the number of actual patients in the trial and the methodology, and explain the limits of the findings with clarity.

Include information about science. Increase your audience's awareness of how science works. Share not only the facts about a new dinosaur fossil discovery but also how paleoscientists know the age of the fossil find. Help people understand how clinical trials for vaccines are conducted and why that matters. There are multiple ways scientists gather and evaluate evidence, and sharing the how, not just the what, develops an appreciation of the scientific enterprise itself. Help your audience understand the wide variety of methods that underpin different fields. Uncertainty is always a part of science, so assist your audience in developing an appreciation of the broad support behind the consensus view. Support their understanding that the weight of the evidence is sufficient to trust settled science. Your audience should walk away with a better understanding of science itself, not just new knowledge of the facts specific to your story.

Provide evidence for scientific claims. Brevity may be the soul of wit, but it is not the foundation of a strong scientific explanation. Be sure to provide your audience with sufficient evidence so that they can evaluate the claims in your communication. Showing the evidence and explaining how scientists reached their conclusions may take up another "inch" in your story, but your audience will not (and should not) trust claims if they cannot see the evidence for themselves.

Know your audience's possible motivations and identities. Group membership is a powerful force. Those who make their living in the Pacific Northwest by logging may push back on a story about overharvesting of old growth forest for economic reasons, and Southern Californians may be leery of a new genetically modified tomato. Political affiliation can serve as an unevaluated perspective on a particular topic, be it resistance to nuclear energy (a concern on the left) or doubting of climate change (a concern on the right).

If you are communicating science to a non-science group, consider using an "ally" from that group. Researchers at Arizona State University have scientists of faith communicate about science to students of faith.[31] Often, in-group members can frame messages to members of their own group better than an outsider.[32] If you are a member of the in-group, leverage your social identity by using inclusive language[33,34]: "We Midwesterners value the family farm, and we know that water must be conserved to conserve the family farm," is a far more effective water conservation message in Nebraska than a visiting environmentalist saying, "Hey, you have to cut back on watering your crops."

Be aware of readers' emotions and attitudes. Science communicators tend to be aware that their topics may be received with strong negative emotions and attitudes. Communications are more effective when they acknowledge the emotions of their audience members directly by lessening the resistance with clear explanations and empowering information. When you write about climate change or pandemics, be sure to include actions individuals can take to mitigate the impacts, such as noting what they can do in their own communities. Science can be frightening, but it can also be uplifting and awe-inspiring, so emphasize the positive when there are legitimate reasons to be positive. When a new treatment is limited in effectiveness but helpful to some, leave your audience with the hopeful story of those whose symptoms were relieved, while tempering how helpful it may be for others. Sharing the wondrous aspects of science can engage your audience, and it may provide some much needed balance for negatives that are overemphasized. Of course, accuracy, not spin, is always the top priority for any science communication.

What Policy Makers Can Do

Policy makers are in a unique role with potential for significant impact on public understanding of science. Policy makers who are in the social or educational arena or those who work in nongovernmental organizations, think tanks, and philanthropic organizations all have a role to play in science policy formation, communication, and evaluation. Individuals who work in these spaces can support public understanding of science and can confront misconceptions and misinformation with education and communication campaigns.

Politicians and those in education policy can have a direct impact on public funding for science, curriculum, teacher professional development, and communication of science and science policy. Those in the policy space know that the workforce needs STEM (science, technology, engineering, and mathematics) employees, but this need cannot be met without investments of public funding and support for science education. This investment is needed at all levels from K–12 through post-secondary and graduate education. There is a significant return on investment in education in general, and in STEM education in particular.[35] However, constituencies will not support science education and research and development (R&D) if they misunderstand and mistrust science. Science funded by corporations is not necessarily biased but does rightfully garner more skepticism, which undermines public trust in the outcomes. It is only through publicly funded science education and R&D that the public has ownership and investment in science.

Decisions on matters of public policy are complex and must be based on a number of factors such as community needs and resources and pragmatics of policy implementation. However, policy makers can and should use science to aid in their decision-making. It has been quipped that every disaster movie begins with a scene where a scientist is being ignored. Those countries that quickly enacted recommendations for science-based practices to limit the spread of COVID-19 were more effective in "flattening the curve." It has been estimated that millions of lives in those countries may have been saved by enacting science-based public health policies.[36]

At the local level, policy makers can have a direct impact on their community's engagement with science. Support for science education in public schools and universities as well as museums, zoos, and community settings benefits the local community. There are local economic benefits to a STEM-educated and -literate community[37] which draws industries that rely on such employees. Science-literate communities have the opportunity to benefit from science knowledge to improve their health and to sustain their communities.

Conclusions

Greater appreciation and understanding of science bring rewards for a healthier and happier life and a more sustainable environment. Each individual can make a difference by being open to scientific ideas and willing to adopt a scientific attitude. Educators can improve their own understanding of the psychology of science appreciation and can support their students toward a greater understanding of science. Science communicators are poised to have an even greater impact as scientists join their ranks and universities value efforts to communicate science to the general public. Policy makers are in the position to push policy levers that include funding science education, using science as an aid to decision-making, and communication efforts in both formal and informal spaces. Understanding the psychological explanations for science denial, doubt, and resistance can offer a foundation for better thinking, education, policy making, and communication about science.

Notes

1. John Zarocostas, "How to Fight an Infodemic," *The Lancet* 395, no. 10225 (2020).
2. Lee McIntyre, *Post-Truth* (Cambridge, MA: MIT Press, 2018).
3. Maxwell T. Boykoff and Jules M. Boykoff, "Balance as Bias: Global Warming and the US Prestige Press," *Global Environmental Change* 14, no. 2 (2004).
4. Bill McKibben, *Falter: Has the Human Game Begun to Play Itself Out?* (New York: Henry Holt and Company, 2019).

5. Lee McIntyre, *The Scientific Attitude: Defending Science from Denial, Fraud, and Pseudoscience* (Cambridge, MA: MIT Press, 2019).

6. Daniel Kahneman, *Thinking, Fast and Slow* (New York: Farrar, Straus and Giroux, 2011).

7. Barbara K. Hofer, "Epistemic Cognition as a Psychological Construct: Advancements and Challenges," in *Handbook of Epistemic Cognition*, ed. Jeffrey Alan Greene, William A. Sandoval, and Ivar Bråten (New York and London: Routledge, 2016).

8. Gale M. Sinatra and Barbara K. Hofer, "Public Understanding of Science: Policy and Educational Implications," *Policy Insights from the Behavioral and Brain Sciences* 3, no. 2 (2016).

9. Dan M. Kahan, "Ideology, Motivated Reasoning, and Cognitive Reflection: An Experimental Study," *Judgment and Decision Making* 8 (2012).

10. Jan Hoffman, "How Anti-Vaccine Sentiment Took Hold in the United States," *New York Times*, September 23, 2019.

11. Jeffrey A. Greene and Brian M. Cartiff, "Think Critically Before Thinking Critically," *Psychology Today*, February 11, 2020, https://www.psychologytoday.com/us/blog/psyched/202002/think-critically-thinking-critically

12. Jessica Lander, "Digital Literacy for Digital Natives," *Usable Knowledge*, January 17, 2018, https://www.gse.harvard.edu/news/uk/18/01/digital-literacy-digital-natives.

13. Samuel S. Wineburg and Sarah McGrew, "Lateral Reading: Reading Less and Learning More When Evaluating Digital Information" (working paper 2017-A1, Stanford University, Stanford, CA, October 6, 2017), http://dx.doi.org/10.2139/ssrn.3048994.

14. Rainie and Janna Anderson. "Code-Dependent: Pros and Cons of the Algorithm Age," PEW Research Center, February 8, 2017, https://www.pewresearch.org/internet/2017/02/08/code-dependent-pros-and-cons-of-the-algorithm-age/.

15. "How Global Warming Works," https://www.howglobalwarmingworks.org.

16. Philipp Schmid and Cornelia Betsch, "Effective Strategies for Rebutting Science Denialism in Public Discussions," *Nature: Human Behaviour* 1 (2019).

17. Katie L. Burke, "8 Myths About Public Understanding of Science," *American Scientist*, February 9, 2015.

18. Katharine Hayhoe, "The Most Important Thing You Can Do to Fight Climate Change: Talk About It" (filmed November 2018 at TEDWomen conference, Palm Springs, CA, TED video, 17:04), https://www.ted.com/talks/katharine_hayhoe_the_most_important_thing_you_can_do_to_fight_climate_change_talk_about_it.

19. Joanna Colwell, "I Didn't Vaccinate My Child and Then I Did," *Addison County Independent*, March 11, 2015.

20. Thomas J. Doherty and Susan Clayton, "The Psychological Impacts of Global Climate Change," *American Psychologist* 66, no. 4 (2011).

21. Gale M. Sinatra and Doug Lombardi, "Evaluating Sources of Scientific Evidence and Claims in the Post-Truth Era May Require Reappraising Plausibility Judgments," *Educational Psychologist* (2020), https://doi.org/10.1080/00461520.2020.1730181.

22. NGSS Lead States, *Next Generation Science Standards: For States, by States* (Washington, DC: National Academies Press, 2013).

23. McIntyre, *The Scientific Attitude*.

24. John A. Bargh and Tanya L. Chartrand, "The Unbearable Automaticity of Being," *American Psychologist* 54, no. 7 (1999).

25. Bill Bishop, *The Big Sort: Why the Clustering of Like-Minded America Is Tearing Us Apart* (New York: Houghton Mifflin Harcourt, 2009).

26. Doug Lombardi, "The Curious Construct of Active Learning," *Psychological Science in the Public Interest* (in press).

27. Laignee Barron, "Here's What the EPA's Website Looks Like After a Year of Climate Change Censorship," *Time*, March 2, 2108, https://time.com/5075265/epa-website-climate-change-censorship/.

28. Shawn Otto, *The War on Science: Who's Waging It, Why It Matters, What We Can Do About It* (Minneapolis: Milkweed, 2016), 24.

29. Lori J. Schroth, "Researchers Create Embryonic Stem Cells without Embryo," *Harvard Gazette*, January 29, 2014, https://news.harvard.edu/gazette/story/2014/01/researchers-create-embryonic-stem-cells-without-embryo/.

30. Ullrich K. H. Ecker, Joshua L. Hogan, and Stephan Lewandowsky, "Reminders and Repetition of Misinformation: Helping or Hindering Its Retraction?," *Journal of Applied Research in Memory and Cognition* 6, no. 2 (2017).

31. M. Elizabeth Barnes and Sara E. Brownell, "Experiences and Practices of Evolution Instructors at Christian Universities That Can Inform Culturally Competent Evolution Education," *Science Education* 102, no. 1 (2018).

32. Viviane Seyranian, "Public Interest Communication: A Social Psychological Perspective," *Journal of Public Interest Communication* 1 (2017).

33. Viviane Seyranian, "Social Identity Framing: A Strategy of Social Influence for Social Change," in *Leader Interpersonal and Influence Skills: The Soft Skills of Leadership*, ed. R. E. Riggio and S. J. Tan (New York: Taylor and Francis, 2013).

34. Seyranian, V., Sinatra, G. M., and Polikoff, M. S. "Comparing communication strategies for reducing residential water consumption," *Journal of Environmental Psychology* 41 (2015), 81–90.

35. Sheila Campbell and Chad Shirley, "Estimating the Long-Term Effects of Federal R&D Spending: CBO's Current Approach and Research Needs," Congressional Budget Office, June 21, 2019, https://www.cbo.gov/publication/54089.

36. Joel Achenback and Laura Meckler, "Shutdowns Prevented 60 Million Coronavirus Infections in the U.S., Study Finds," *Washington Post*, June 8, 2020. https://www.washingtonpost.com/health/2020/06/08/shutdowns-prevented-60-million-coronavirus-infections-us-study-finds/.

37. Joel Mokyr, "Building Taller Ladders: Technology and Science Reinforce Each Other to Take the Global Economy Ever Higher," *Finance & Development*, 55, no. 2 (2018).

References

Achenback, Joel, and Laura Meckler. "Shutdowns Prevented 60 Million Coronavirus Infections in the U.S., Study Finds." *Washington Post*, June 8, 2020. https://www.washingtonpost.com/health/2020/06/08/shutdowns-prevented-60-million-coronavirus-infections-us-study-finds/.

Bargh, John A., and Tanya L. Chartrand. "The Unbearable Automaticity of Being." *American Psychologist* 54, no. 7 (1999): 462–79.

Barnes, M. Elizabeth, and Sara E. Brownell. "Experiences and Practices of Evolution Instructors at Christian Universities That Can Inform Culturally Competent Evolution Education." *Science Education* 102, no. 1 (2018): 36–59.

Barron, Laignee. "Here's What the EPA's Website Looks Like After a Year of Climate Change Censorship." *Time*, March 2, 2108. https://time.com/5075265/epa-website-climate-change-censorship/.

Bishop, Bill. *The Big Sort: Why the Clustering of Like-Minded America Is Tearing Us Apart.* New York: Houghton Mifflin Harcourt, 2009.

Boykoff, Maxwell T., and Jules M. Boykoff. "Balance as Bias: Global Warming and the US Prestige Press." *Global Environmental Change* 14, no. 2 (2004): 125–36.

Burke, Katie L. "8 Myths About Public Understanding of Science." *American Scientist*, February 9, 2015.

Campbell, Sheila, and Chad Shirley. "Estimating the Long-Term Effects of Federal R&D Spending: CBO's Current Approach and Research Needs." Congressional Budget Office, June 21, 2019. https://www.cbo.gov/publication/54089.

Colwell, Joanna. "I Didn't Vaccinate My Child and Then I Did." *Addison County Independent*, March 11, 2015.

Doherty, Thomas J., and Susan Clayton. "The Psychological Impacts of Global Climate Change." *American Psychologist* 66, no. 4 (2011): 265–76.

Ecker, Ullrich K. H., Joshua L. Hogan, and Stephan Lewandowsky. "Reminders and Repetition of Misinformation: Helping or Hindering Its Retraction?" *Journal of Applied Research in Memory and Cognition* 6, no. 2 (2017): 185–92.

Greene, Jeffrey A., and Brian M. Cartiff. "Think Critically Before Thinking Critically." *Psychology Today*, February 11, 2020, https://www.psychologytoday.com/us/blog/psyched/202002/think-critically-thinking-critically.

Hayhoe, Katharine. "The Most Important Thing You Can Do to Fight Climate Change: Talk About It." Filmed November 2018 at TEDWomen conference, Palm Springs, CA. TED video, 17:04. https://www.ted.com/talks/katharine_hayhoe_the_most_important_thing_you_can_do_to_fight_climate_change_talk_about_it.

Hofer, Barbara K. "Epistemic Cognition as a Psychological Construct: Advancements and Challenges." In *Handbook of Epistemic Cognition*, edited by Jeffrey Alan Greene, William A. Sandoval, and Ivar Bråten, 19–38. New York and London: Routledge, 2016.

Hoffman, Jan. "How Anti-Vaccine Sentiment Took Hold in the United States." *New York Times*, September 23, 2019.

"How Global Warming Works." https://www.howglobalwarmingworks.org.

Kahan, Dan M. "Ideology, Motivated Reasoning, and Cognitive Reflection: An Experimental Study." *Judgment and Decision Making* 8 (2012): 407–24.

Kahneman, Daniel. *Thinking, Fast and Slow.* New York: Farrar, Straus and Giroux, 2011.

Lander, Jessica. "Digital Literacy for Digital Natives." *Usable Knowledge*, January 17, 2018. https://www.gse.harvard.edu/news/uk/18/01/digital-literacy-digital-natives.

Lombardi, Doug, Shipley, T. F., Astronomy Team (Bailey, J. M, Bretones, P. S., Prather, E. E.), Biology Team (Ballen, C. J., Knight, J. K., Smith, M. K.), Chemistry Team (Stowe, R. L., Cooper, M. M.), Engineering Team (Prince, M.), Geography Team (Atit, K., Uttal, D. H.), Geoscience Team (LaDue, N. D., McNeal, P. M., Ryker, K., St. John, K., van der Hoeven Kraft, K. J.), & Physics Team (Docktor, J. L.). "The Curious Construct of Active Learning." *Psychological Science in the Public Interest* 22 (1) (2021). https://doi.org/10.1177/1529100620973974

McIntyre, Lee. *Post-Truth*. Cambridge, MA: MIT Press, 2018.

McIntyre, Lee. *The Scientific Attitude: Defending Science from Denial, Fraud, and Pseudoscience*. Cambridge, MA: MIT Press, 2019.

McKibben, Bill. *Falter: Has the Human Game Begun to Play Itself Out?* New York: Henry Holt and Company, 2019.

Mokyr, Joel. "Building Taller Ladders: Technology and Science Reinforce Each Other to Take the Global Economy Ever Higher." *Finance & Development* 55, no. 2 (2018): 32–35.

NGSS Lead States. *Next Generation Science Standards: For States, by States*. Washington, DC: National Academies Press, 2013.

Otto, Shawn. *The War on Science: Who's Waging It, Why It Matters, What We Can Do About It*. Minneapolis: Milkweed, 2016.

Rainie, Lee, and Janna Anderson. "Code-Dependent: Pros and Cons of the Algorithm Age." PEW Research Center, February 8, 2017. https://www.pewresearch.org/internet/2017/02/08/code-dependent-pros-and-cons-of-the-algorithm-age/.

Schmid, Philipp, and Cornelia Betsch. "Effective Strategies for Rebutting Science Denialism in Public Discussions." *Nature: Human Behaviour* 3, no. 9 (2019): 931–39.

Schroth, Lori J. "Researchers Create Embryonic Stem Cells without Embryo." *Harvard Gazette*, January 29, 2014. https://news.harvard.edu/gazette/story/2014/01/researchers-create-embryonic-stem-cells-without-embryo/.

Seyranian, Viviane. "Public Interest Communication: A Social Psychological Perspective." *Journal of Public Interest Communication* 1 (2017): 57–77.

Seyranian, Viviane. "Social Identity Framing: A Strategy of Social Influence for Social Change." In *Leader Interpersonal and Influence Skills: The Soft Skills of Leadership*, edited by R. E. Riggio and S. J. Tan, 207–42. New York: Taylor and Francis, 2013.

Seyranian, V., Sinatra, G. M., and Polikoff, M. S. "Comparing communication strategies for reducing residential water consumption," *Journal of Environmental Psychology* 41 (2015), 81–90.

Sinatra, Gale M., and Barbara K. Hofer. "Public Understanding of Science: Policy and Educational Implications." *Policy Insights from the Behavioral and Brain Sciences* 3, no. 2 (2016): 245–53.

Sinatra, Gale M., and Doug Lombardi. "Evaluating Sources of Scientific Evidence and Claims in the Post-Truth Era May Require Reappraising Plausibility Judgments." *Educational Psychologist* 55, no. 3 (2020): 120–31. https://doi.org/10.1080/00461520.2020.1730181.

Wineburg, Samuel S., and Sarah McGrew. "Lateral Reading: Reading Less and Learning More When Evaluating Digital Information." Working paper 2017-A1, Stanford University, Stanford, CA, October 9, 2017. http://dx.doi.org/10.2139/ssrn.3048994.

Zarocostas, John. "How to Fight an Infodemic." *The Lancet* 395, no. 10225 (2020): 676.

Index

For the benefit of digital users, indexed terms that span two pages (e.g., 52–53) may, on occasion, appear on only one of those pages.

absolutism, 99–102
academic emotion, 151–52
accuracy goals, 126
acid rain, 11–13
action steps
    dealing with cognitive biases
        for educators, 91–92
        for individuals, 90–91
    supporting digital literacy
        for educators, 41–42
        for individuals, 39–41
        for policy makers, 43–44
        for science communicators, 42–43
    supporting science education
        for educators, 64
        for policy makers, 65
        for science communicators, 64–65
    understanding beliefs about knowledge
            and knowing
        for educators, 114–15
        for individuals, 112–16
        for science communicators, 115–16
    understanding emotions and attitudes
        for educators, 154–55
        for individuals, 154
        for informal learning
            environments, 155
        for science communicators, 155–56
    understanding motivated reasoning
        for individuals, 135–36
        for science communicators, 136–37
    See also field guide to address
            science denial
African Americans, 6–7
algorithmic literacy, 35, 40–41, 42, 166–67
algorithms
    definition of, 33–34, 40–41

effect on online searches, 33–35, 43,
        162, 166–67
    role in confirmation bias, 87
Ariely, Dan, 78
Armor, David, 127–28
artificial intelligence, 35, 166–67. *See also*
        algorithmic literacy; algorithms
attitudes
    belief-based attitude, 9
    motivated reasoning and, 125–28
    scientific attitude, 8, 17, 105–6, 107,
        149, 162
    social identity and, 129–34, 163–64
    See also emotions and attitudes
availability heuristic, 88, 162–63

Baehr, Jason, 115
balance as bias, 102–5
Barnes, Elizabeth, 134
belief-based attitude, 9
beliefs about knowledge and knowing
    action steps
        for educators, 114–15
        for individuals, 112–16
        for science communicators, 115–16
    dealing with new information and
            conflicting ideas, 97–99
    epistemic cognition, 99–102
    epistemic trust in science, 109–12
    misconceptions about science as a way
            of knowing, 106–9
    multiplistic thinking, 102–5
    scientific literacy and the practice of
            science, 105–6
benevolence, 27–28, 113
bias
    balance as bias, 102–5

bias (*cont.*)
　finding unbiased information
　　sources, 29–30
　unbiased rationality, 125
　See also cognitive biases
Big Sort: Why the clustering of Like-Minded
　　America Is Tearing us Apart, The
　　(Bishop), 130–31
Bishop, Bill, 130–31
bounded understanding of science, 26
Bromme, Rainer, 43
Broughton, Suzanne, 143
Brownell, Sarah, 134

cafeteria denial, 9
Carr, Nicholas, 35–36, 86
Carson, Rachel, 11–13
causal links, 127–28
Chinn, Clark, 58–59
Clark, Dav, 85–86
climate anxiety, 148
climate change
　belief-based attitude toward, 9
　confidence by coherence and, 80–81
　confirmation bias and, 32–33
　consensus opinion on, 108
　DuPont's dismissal of, 11–12
　gap between scientific knowledge and
　　public understanding, 3–4
　as an "inconvenient truth," 13
　informational bias and, 103
　lack of understanding
　　surrounding, 12–13
　potential for long-term catastrophe, 12
cognitive biases
　action steps for educators, 91–92
　action steps for individuals, 90–91
　anecdotal versus scientific
　　thinking, 84–85
　availability heuristic, 88, 162–63
　confidence by coherence, 80–81
　confirmation bias, 87–88
　illusion of understanding, 85–86
　intuitive theories, 81–84
　as key psychological construct, 162–63
　role in decision-making, 78
　thinking dispositions, 88–90
cognitive incongruity, 152–53

cognitive shortcuts, 32–33
coherence, 80–81, 91, 166
communicators. *See* science
　communicators
complacent ignorance, 86
conceptual change, 59–62
confidence by coherence, 80–81, 91, 166
confirmation bias, 32–33, 87–88
Conway, Erik, 10, 11, 106
Coronavirus. *See* COVID-19
COVID-19
　belief-based attitude toward, 9
　confirmation bias and, 32–33
　filter bubbles and echo chambers, 33–35
　slow response to, 13
　variation in value of expertise
　　concerning, 29
creationism, 13, 36–37
Crick, Francis, 11
critical thinking
　developing skills in, 39–40, 41, 152, 166
　need for, 29–30
　pro-con format and, 104–5
　risk of foreclosure and, 154
　scientific versus anecdotal
　　thinking, 84–85
　social groups and, 132
　tackling cognitive biases
　　through, 162–63
　thinking dispositions and, 88–90
　training in digital literacy, 36–39
crowdsourcing, 131–32

Darwin, Charles, 10–11
Death of Expertise, The (Nichols), 28–29, 101
denial. *See* science denial
denialism, 6. *See also* science denial
digital literacy
　action steps supporting, 39–44
　bounded understanding of science, 26
　confirmation bias and, 32–33
　degree of worldwide digital
　　connectivity, 24–26
　developing, 36–39
　evaluating science information
　　online, 26–28
　expertise versus experience/personal
　　testimony, 31–32

"Google knowing" versus
        understanding, 35–36
    hidden structure of online
        searching, 33–35
    science claims found online, 23
    valuing and decline of expertise, 28–30
digitalliteracy.gov, 38
disinformation
    definition of, 25–26
    as indiscernible from
        information, 161–62
    skepticism and mistrust created by, 8
    spread for corporate gain, 8
    spread online, 129
    tackling with digital literacy, 172–73
diversification, 6–7
diversity of belief, 101–2
doubt. *See* science doubt

echo chambers, 13, 33–35
eco-anxiety, 148
educators
    action steps for
        beliefs about knowledge and
            knowing, 114–15
        cognitive bias, 91–92
        digital literacy, 38, 41–42
        science education, 64
        understanding emotions and
            attitudes, 154–55
    Next Generation Science Standards
        (NGSS), 55–59
    role in overcoming science
        denial, 17–19
    teacher preparation in science, 53–54
    undermining acceptance of expert
        knowledge, 104–5
    See also science education
emotional thought, 145–46
emotional valence, 147
emotions and attitudes
    action steps
        for educators, 154–55
        for individuals, 154
        for informal learning
            environments, 155
        for science communicators, 155–56
    attitudes versus emotions, 148–50

    connection of thoughts to
        emotions, 145–46
    influence on science
        understanding, 143–45
    as key psychological construct, 164
    in learning of science in school, 151–53
    in public learning of science, 150–51
    role of emotions in thinking about
        science, 146–48
    understanding science through
        emotions, 142–43
    variety of emotions toward science, 145
Enlightenment Now (Pinker), 5
epistemic cognition, 15–16, 98–102,
        116, 163
epistemic competence, 115
epistemic trust, 109–11, 163
epistemic trust in science, 109–12
epistemic vigilance, 113, 115
epistemic virtues, 115
epistemology of science, 105
essentialist thinking, 10–11
evaluating information/scientific
        claims online
    digital literacy, 36–39
    evaluating science information
        online, 26–28
    expertise versus experience/personal
        testimony, 31–32
    valuing and decline of expertise, 28–30
evaluativism, 99–102, 115
evidence
    availability heuristic and, 88, 162–63
    bounded understanding of science
        and, 26
    caring about evidence, 113, 114, 149
    confidence by coherence, 80–81
    confirmation bias and, 87–88
    critical evaluation of, 4, 26–28, 38–39,
        40, 50–51, 64
    epistemic cognition and, 99–102
    evidence-evaluation activities, 58–59
    false equivalencies and, 102–5
    falsification and disregard of, 10, 123
    motivated reasoning and, 124–29
    providing context for emerging, 64–65
    providing for scientific claims, 115
    refutation texts and, 60, 83–84

evidence (*cont.*)
  rejection of, 13–14
  scientific attitude and, 5–6, 8, 106, 107
  seeing patterns in data as, 31–32
  seeking and evaluating, 9–10, 90–91
  social identity and, 132–34
  System 1 and System 2 thinking, 79–80
  thinking dispositions and, 88–90
  weight of evidence reporting, 102–3
  See also scientific evidence
evidence-based decisions, 154, 164–65
evidence-based knowledge, 113, 163
evolution theory
  conflicts with social groups, 133–34
  dinosaurs and humans, 36–37
  epistemic cognition and, 99–102
  misconceptions concerning, 107
  resistance to fueled by essentialist
    thinking, 10–11
  teaching alongside creationism, 13
expertise
  balance as bias, 102–5
  celebrity opinions, 115–16
  evaluating science information
    online, 26–28
  versus experience and personal
    testimony, 31–32
  from in-group members, 134
  post-truth messaging and, 101
  valuing and decline of expertise, 28–30
extreme relativism, 101–2

Facebook
  expertise versus experience/personal
    testimony, 31–32
  filter bubbles and echo
    chambers, 33–35
  social identity and, 55, 163–64
  valuing and decline of expertise, 28–30
  See also social media
false clarity, 108
false equivalencies, 102–5
Fernbach, Philip, 86
field guide to address science denial
  communicating with others, 169–70
  educators, 171–75
  goals of, 164–65
  individuals, 165–68

policy makers, 179–80
  science communicators, 175–79
  See also action steps
filter bubbles, 33–35, 87
Fiske, Susan, 79
foreclosure, 154
Franklin, Rosalind, 11
functional skepticism, 9–10
funding disclosures, 29–30

Galileo, 10–11, 161
genetically modified organisms
    (GMOs), 60–61
germ theory, 11
Global Climate Coalition, 12–13
global warming. *See* climate change
Goldenburg, Maya, 127–28
Google
  confirmation bias and, 32–33
  effect of algorithms on Google
    searches, 33–35
  evaluating science information
    online, 26–28
  "Google knowing" versus
    understanding, 35–36
  number of Google searches per
    minute, 25
Gore, Al, 13
Grajal, Alejandro, 151
Grasswick, Heidi, 112
*Greatest Hoax: How the Global Warming
    Conspiracy Threatens Your Future,
    The* (Inhofe), 80–81
group affiliation, 130

Hansen, James, 12–13
Hendriks, Fredrike, 112
heuristics, 26–27, 88, 162–63
HowGlobalWarmingWorks.org, 85–86
hypotheses, versus theories, 107

identity. *See* social identity
illusion of objectivity, 127–28
illusion of understanding, 85–86, 162–
    63, 166
Immordino-Yang, Mary Helen, 145–46
impartiality, 102–5
"inconvenient truth," 13

individual action steps
  cognitive biases, 90–91
  dealing with motivated
    reasoning, 135–36
  digital literacy, 39–41
  for a post-truth age of
    manipulation, 13–14
  understanding beliefs about knowledge
    and knowing for, 112–16
  understanding emotions and
    attitudes, 154
individual rights, 101–2
informal learning environments
  as invaluable learning
    opportunities, 150–51
  NGSS instruction in, 59
  understanding emotions and attitudes
    action steps, 155
information literacy
  challenges of teaching, 36–39
  definition of, 39–40
  fostering in students, 41–42
  influence of search algorithms
    on, 33–35
  websites teaching, 38, 166
  See also digital literacy
ingroups, 129–34, 178
Inhofe, Jim, 80–81, 88
integrity, 27–28
"intelligent design," 13
internet. *See* digital literacy; Google;
  online information
intuitive theories, 78, 81–84, 85, 91

journalists
  contribution to social construction of
    ignorance, 103
  misleading headlines by, 108
  problem of false equivalency, 103
  weight of evidence reporting, 102–3
  See also science communicators

Kahan, Dan M., 123–24
Kahneman, Daniel, 26–27, 78–80, 81
knowing versus understanding, 35–36
knowledge. *See* beliefs about knowledge
  and knowing
knowledge deficit view, 4, 54–55, 63

*Knowledge Illusion: Why We Never Think
  Alone* (Sloman and Fernbach), 86
Koehler, Derek, 103–4
Kunda, Ziva, 125, 127

learning
  informal learning environments, 59,
    150–51, 155, 173–74
  machine learning, 35
  motivated reasoning and, 128–29
  role of emotions and attitudes in
    public, 150–51
  role of emotions and attitudes in
    school, 151–53
  See also science learning
LeGrand, Homer, 128
Lodge, Milton, 125–26
Lombardi, Doug, 58, 129
Lord, Charles G., 125–26
Lynch, Michael, 32–33, 35–36

machine learning, 35. *See also* algorithmic
  literacy; algorithms
Mann, Michael, 6, 63
McCarthy, Jenny, 32–33
McIntyre, Lee, 5–6, 8, 9, 107, 134, 149
McKibben, Bill, 12
media literacy, 38. *See also* digital literacy
Merchants of Doubt (Oreskes and
  Conway), 10, 11
Mercier, Hugo, 87
misconceptions
  about science as a way of
    knowing, 106–9
  addressing, 65–66
  concerning uncertainty in science, 107–
    8, 176–77
  confronting through refutation texts,
    60–61, 83–84, 91, 101–2
  versus intuitive theories, 82
misinformation
  definition of, 25–26
  versus intuitive theories, 82
  multiplication of on social media, 33–35
motivated reasoning
  action steps
    for individuals, 135–36
    for science communicators, 136–37

motivated reasoning (*cont.*)
  influence on decision making, 125–28
  as key psychological construct, 163
  social identity and, 129–34
  when learning about science
    online, 128–29
motivation, types of, 122–25
multiplism, 99–103
multiplistic thinking, 101, 102–5
museums. *See* informal learning
    environments
myside bias, 87

need for cognition, 89–90
Next Generation Science Standards
    (NGSS), 55–59, 105–6, 162
NGSS. *See* Next Generation Science
    Standards
Nichols, Tom, 28–29, 101
Nisbett, Richard, 31–32

online information
  evaluating science information
    online, 26–28
  expertise versus experience/personal
    testimony, 31–32
  hidden structure of online
    searching, 33–35
  motivated reasoning and, 128–29
  See also digital literacy
open-minded thinking, 89
Oerskes, Naomi, 3, 5–6, 10, 11, 106, 128
Origin of the Species (Darwin), 10–11
out-groups, 129–31, 133, 134, 136
ozone depletion, 11–13

pandemics
  addressing with vaccines, 56
  changing recommendations
    during, 176–77
  cultivating scientific attitudes
    during, 165
  economic concerns during, 78–79, 126
  halting spread of, 13, 25
  misinformation and disinformation
    during, 129
  role of emotions in thinking about
    science, 172
  role of political leaders in, 88
  science communicator action
    steps, 178–79
  scientific inquiry and, 106
  skepticism and mistrust during, 8, 9,
    81, 161
  social distancing during, 51
  spread of fueled by denial of, ix
  variation in value of expertise
    concerning, 29, 44
  wearing masks to prevent contagion of,
    3, 55, 122–23, 132, 154
  See also COVID-19
Paris Climate Accord, 122–23
peer review, 6, 9–10, 26, 106, 111, 112,
    113–14, 173
personal testimony, 31–32
pesticides, 11–13
Pew Science Knowledge Quiz, 51
Pinker, Steven, 5
Pinterest, 43–44
PISA. *See* Programme for International
    Student Assessment
plate tectonics, 128
*Plate Tectonics: An Insider's History of
    the Modern Theory of the Earth*
    (Oreskes and LeGrand), 128
plausibility, 27–28, 127, 132
plutoed, 143
*Pluto File: The Rise and Fall of America's
    Favorite Planet, The* (Tyson), 143
polarization, 133
policy
  effect of conflicting opinions on, 104
  effect of science denial on, 9–10
  eliminating politics from, 19
  guiding with scientific evidence, 6–
    7, 164–65
  public understanding of science and, 6
  trust in scientific authorities and, 110
policy makers
  combating science denial, 179–80
  digital literacy action steps, 43–44
  evaluating online
    information, 38–39
  manipulated by corporations, 11–12
  role in overcoming science
    denial, 17–19

science-based decision making
  by, 164–65
science education action steps, 65
selecting, 168
social identity and, 129–30
valuing scientists and data, 8,
  18–19
wariness of limited findings, 29–30
"post-truth" messaging
  devaluing of truth in, 13–14
  fostering multiplism through, 101
  resisting, 113, 162, 167
predictably irrational thinking, 78
ProCon.org, 104–5
Programme for International Student
  Assessment (PISA), 52–53
proximity in time argument, 127–28
psychological constructs, x–xi, 14–
  17, 162–64
public understanding of science
  bounded understanding, 26
  emotions and attitudes in public
    learning of science, 150–51
  factors influencing scientific
    understanding, 62–63
  gap between scientific knowledge and
    public understanding, 3–4
  "Google knowing" versus
    understanding, 35–36
  scientific literacy around the
    world, 52–54
  scientific literacy in the US, 51–52
  understanding science through
    emotions, 142–43
  See also science education
"Public Understanding of Science: Policy
  and Educational Implications"
  (Sinatra and Hofer), xi

racism, 6–7
Ranney, Michael, 85–86
reasoning. See motivated reasoning
refutation texts, 60–61, 83–84, 91,
  101–2
relativism, 101–2
religious groups, 131
  roots of, 6
  to scientific evidence, 13–14, 161–62

understanding points of
  potential, 18–19
Ross, Lee, 31–32

Scienceblind: Why Our Intuitive Theories
  About the World Are So Often
  Wrong (Shtulman), 82
science communication, 16–17, 18, 64–65,
  175–76, 178–79
science communicators
  action steps for
    beliefs about knowledge and
      knowing, 115–16
    dealing with motivated
      reasoning, 136–37
    digital literacy, 42–43
    science education, 64–65
    understanding emotions and
      attitudes, 155–56
  role in overcoming science denial, 17–19
  See also journalists
science denial
  in current day, 13–14
  definition of terms, 9–10
  fallibility of science, 6–7
  gap between scientific knowledge and
    public understanding, 3–4
  history of, 10–11
  in the modern era, 11–13
  overcoming
    challenges of, 161–62
    field guide for, 164–80
    key psychological constructs, 162–64
    roles of educators, communicators,
      scientists, and policy
      makers, 17–19
  psychological explanations for, 14–17
  public understanding of science, 8
  value of science, 4–6
science doubt
  definition of, 10
  fostering for corporate gain, 11–13
  gap between scientific knowledge and
    public understanding, 3–4
  history of, 10–11
  in modern era, 11–13
  politicization of, 13
  psychological issues fostering, 14–17

science education
    action steps supporting, 63–65
    conceptual change and, 59–62
    factors influencing scientific
        understanding, 62–63
    knowledge deficit view, 54–55
    Next Generation Science
        Standards, 55–59
    role in decision-making, 50–51
    scientific literacy around the
        world, 52–54
    scientific literacy in the US, 51–52
science knowledge
    beliefs about knowledge and
        knowing, 98
    benefitting from, 180
    epistemic trust and, 110
    ever-evolving nature of, 107
    individual action steps, 114, 169
    key psychological constructs and, x–xi
    prioritizing over intuition, 85
    role of educators in enhancing, 58, 171
    state of in the US, 52–54
    versus thinking like a scientist,
        55, 58–59
science learning
    active science instruction, 57–58
    augmented and virtual reality and, 59
    fostering openness to, 164
    improving support for, 175–79
    intuitive theories and, 83
    See also learning
Science resistance
    definition of, 9–10
    field guide to addressing, 164–80
    fueled by essentialist thinking, 10–11
    growing problem of, ix
    history of, 10–11
    key psychological reasons for, 14–17
science skepticism, 9–11
science writers. See science
        communicators
scientific attitude
    adopting and maintaining, 8
    caring about evidence, 149
    fostering in students, 17, 162
    four basic premises of, 105–6
    summation of, 107

scientific evidence
    versus belief-based attitudes, 9
    confirming or denying message
        through, 166
    encouraging students to value, 81
    guiding others to accept, 170
    guiding policy decisions, 164–65
    versus intuitive theories, 83–84
    judging validity of online
        material, 172–73
    misguided attempts at fairness and, 177
    resistance to, 161–62
    scientific versus anecdotal
        thinking, 84–85
    social groups and, 133–34
    teaching value of, 114
    valuing, 113
scientific expertise
    expertise versus experience/personal
        testimony, 31–32
    valuing and decline of expertise, 28–30
scientific literacy, 52–54, 105–6, 142–43
scientific method, 5–6, 55–56, 106, 108–9,
        114, 173
scientific thinking
    anecdotal versus scientific
        thinking, 84–85
    definition of, 107
    fostering in students, 174–75
    nurturing value of, 168
    System 1 and System 2 thinking, 79–80
scientists
    basic premises shared by, 105–6
    degree of confidence in, 29–30
    role in overcoming science
        denial, 17–19
search for truth, 113
self-observation, 108–9
self-regulation, 37–38
Shallows: What the Internet Is Doing to Our
        Brains, The (Carr), 35–36, 86
Shell, 12–13
Shtulman, Andrew, 82, 84, 91
Silent Spring (Carson), 11–13
skepticism
    history of science skepticism, 10–11
    need for science skepticism, 9–10
Sloman, Steve, 86

Slovic, Paul, 87–88
smoking, 11–13
social identity, 55, 129–34, 162, 163–
        64, 178
social media
    algorithms determining feeds on, 65–
        66, 162, 166–67
    amplification of existing beliefs by, 13
    conflicting information found on, 3
    crowdsourcing opinions on, 131–32
    degree of use in US, 24–25
    evaluating claims on, 38–39, 99, 113
    filter bubbles and echo chambers on,
        33–35, 42, 130–31, 132, 174
    influence on interpretation of
        information, 127
    misinformation spread on, 129
    motives impacting, 123
    regulatory policies of, 43–44
    See also Facebook
Speedometry curriculum, 57–58, 155
Sperber, Dan, 87
Stanovich, Keith E., 89
STEM (science, technology, engineering,
        and mathematics), 52, 179
STEM education, 52, 53–54, 65,
        179, 180
Stoknes, Per Espen, 6, 9, 108
System 1 and System 2 thinking, 26–28,
        37–38, 78–80

Taber, Charles, 125–26
Taylor, Shelly, 79
theory, versus hypothesis, 107
thinking dispositions, 88–90, 91
thinking scientifically
    fostering in students, 174
    versus thinking anecdotally, 84–85
    See also scientific thinking
Thunberg, Greta, 101
tolerance, 101–2
trust. See epistemic trust; epistemic trust
        in science
truth
    devaluing of, 13–14
    "inconvenient truth," 13

"post-truth" messaging, 101
    valuing the search for, 113
Truth, The (Wakefield and
        McCarthy), 32–33
Twitter, 43–44
Tyson, Neil Degrasse, 143, 170

unbiased rationality, 125
uncertainty in science
    appreciating consensus views, 177–78
    embraced by scientists, 27
    emphasized to create misinformation,
        11–13, 103
    induced through dissenting expert
        opinions, 104
    misconceptions concerning, 107–
        8, 176–77
    versus tentativeness, 169
    worldview of multiplism and, 100
understanding. See illusion of
        understanding; knowing
        versus understanding; public
        understanding of science

vaccination hesitancy, 13, 108
vaccinations
    confirmation bias and, 32–33
    crowdsourcing opinions on, 132
    erroneous link to autism, 111, 127–28
    finding unbiased information
        sources, 43–44
    gap between scientific knowledge and
        public understanding, 3–4
    role of emotions in thinking about
        science, 149
    science claims found online, 23–24

Wakefield, Andrew, 32–33
Watson, James, 11
Wegener, Alfred, 11, 128
weight of evidence reporting, 102–3
West, Richard F., 89
Why Trust Science (Oreskes), 5–6
Wikipedia, 28–29, 36–37

zoos. See informal learning environments